Bill

£5

FRESHWATER LIFE

Freshwater Life

JOHN CLEGG Hon. F.L.S.

With 16 colour plates, 48 half-tone plates from photographs by the author and 88 figures in the text

FREDERICK WARNE & CO LTD
FREDERICK WARNE & CO INC
London and New York

First published as *Freshwater Life of the British Isles* by
FREDERICK WARNE & CO LTD
LONDON, ENGLAND 1952
Second Edition   1959
Third Edition   1965

Revised and reset
© FREDERICK WARNE & CO LTD
1974

To
GILLIAN and JACQUELINE

ISBN 0 7232 1762 9

Printed in Great Britain by
Butler & Tanner Ltd, Frome and London

680.1073

CONTENTS

LIST OF PLATES

* (*colour plate*)

PREFACE TO THE FOURTH EDITION

Since this book was written twenty-five years ago, changes have taken place both in its type of reader and in the way its subject is studied. The amateur pond-hunters, who as I said in the first Preface, 'indulge in pond-hunting purely for their own pleasure', are fewer but, on the other hand, there are now many more people at school or college whose studies include specific projects on fresh water, or whose examination syllabus includes some aspect of freshwater ecology. More precise information is sought by this kind of reader, but one hopes that their less personally motivated approach to the subject does not result in their visits to freshwater habitats being any less enjoyable, or prevents them from standing back occasionally from their labours to reflect on some particularly beautiful adaptation of structure or behaviour in the organisms they are studying.

The necessity of resetting the type for this fourth edition has given the opportunity that was not practicable in previous new editions of extending and modifying the text fairly extensively to provide the more detailed information that today's students require but without, I hope, making the book less acceptable to readers with no formal biological training who, apparently, have appreciated the simple approach to the subject in the earlier editions. The majority of the line illustrations have been redrawn and a few plates have been changed; colour plates of dragonflies have been added to replace those of less closely studied animals. The book was never intended as an identification manual. It is quite impossible in a volume of this size to provide identification features of more than a tiny fraction of the immense number of species of freshwater plants and animals, and the reader must still make use of keys and other taxonomic works dealing with specific groups, some of which are listed in the Bibliography. The book is primarily intended to be an introduction to the diversity of living organisms in lakes, ponds, streams and rivers, and some of the factors that influence them, and to give brief details of the life-histories of the commonest ones.

Nomenclature, and particularly the frequent changes that are needed to keep it up to date, is still a problem for authors and publishers of works such as this which cover a wide range of plant and animal groups. One feels that the time has come to stabilize internationally accepted and familiar names. In the present edition Lord Rothschild's *A Classification of Living Animals*, second edition, and L. A. Borradaile's *The Invertebrata*, fourth edition, have been used as a framework for the animals, and *Flora of the British Isles*, second edition, by

A. R. Clapham, T. G. Tutin and E. F. Warburg for the flowering plants. I gratefully acknowledge the help I have received from these and the other works of reference detailed in the Bibliography on pages 268 to 272, which have been freely consulted during the writing of this book.

Dr Barbara Gilchrist, Dr T. T. Macan and Dr Barbara Walsh (Mme Maetz) have at various times during the life of the book read through the whole text and made valuable suggestions. I am especially grateful to them and also to the following who have helped with specific points on which they were authorities: Mr G. O. Allen, Dr Vera Collins, Dr Geoffrey Fryer, F.R.S., Dr F. Green-shields, Mr J. Hanley, Dr Anna B. Hastings, Miss Brenda Knudson (Mrs T. Kipling), Mr F. J. H. Mackereth, Dr Erica Swale, Dr J. F. Talling, Dr G. Willoughby and Mr T. Zboinski. Mr John Horne, the Librarian of the Fresh-water Biological Association Library at the Windermere Laboratory, and his staff, have been most assiduous in finding for me difficult references. While expressing my sincere thanks to all these friends and acknowledging their help, I must take personal responsibility for any errors or ambiguities that may still remain. The short quotation from *Inversnaid* by Gerard Manley Hopkins on page 260 is by kind permission of The Rev. Father Bernard Hall, S.J., for the Society of Jesus, the copyright holders.

New line drawings have been made by Mr Gordon Riley and Mrs Joan Worthington. Figs. 42 and 43, top pair, are by Mr E. Hollowday. The colour plates are by the late E. C. Mansell. I thank all these artists for their valuable contribution to the book. Acknowledgments are also due to the Council of the Freshwater Biological Association and the authors concerned for per-mission to adapt Fig. 5 from a diagram that appeared in Scientific Publication No. 11, *Freshwater Biology and Water Supply in Britain*, by W. H. Pearsall, A. C. Gardiner and F. Greenshields; to Dr Hilda Canter (Mrs J. W. G. Lund) for permission to base Fig. 22 on an illustration which appeared in her paper *Studies on Plankton Parasites*: No. 1 (1948); and to Dr Erica Swale for help with Fig. 23 of *Euglena* and *Ceratium*.

The photographs from which the black and white plates have been prepared are my own, but many of them could not have been taken without the generous help in providing material by loan or gift, from the following: Messrs G. O. Allen, J. D. Allonby, J. Barling, E. E. Dennis, J. Hanley, Miss Brenda Knudson, Dr T. T. Macan, Mr E. R. Newmarch, Mr S. Nield, Dr Barbara Walsh, Dr Mary Young, the Councils of the Royal Microscopical Society and of the Quekett Microscopical Club, and Messrs Watkins and Doncaster.

Finally, I should like to express my great appreciation to the publisher's staff for overcoming the many problems that arose in the course of preparing this new edition for press.

August 1973   JOHN CLEGG

I : INTRODUCTION

Until recent times, mankind's dependence on fresh water made it essential for him to live within easy reach of lakes, rivers and other sources of this indispensable substance. Perhaps this is why, even today, when the amenities of modern life have removed the necessity of close proximity to water, the waterside still holds such an attraction and fascination for us as a place of recreation. The child fishing for tadpoles, the angler on the river bank and the scientist searching the depths of a lake with net and dredge are, perhaps, unconsciously obeying an instinct to turn to the water that has been a part of man's nature from the earliest times.

At first man's interest in the water was merely to satisfy his thirst, provide some of his food and a means of cleansing himself, but as time went on his descendants could hardly fail to notice certain aquatic phenomena. Two of the plagues of Egypt were of this kind: the plague of frogs and the rivers that turned to 'blood' (through the presence of algae). Legends of 'will-o'-the-wisps' (the spontaneous ignition of marsh gas); water-maids with long, green hair; magic shrouds (masses of blanket algae) and many more became a part of folklore and were handed down from generation to generation.

Later, more orderly observations came to be made, for example on the habits of fish, the mass emergences of aquatic insects or the covering of the water surface with a scum of algae ('water bloom' or 'the breaking of the meres'). A body of knowledge accumulated in time—a strange mixture of fact and fancy, some of which became recorded in early writings and was thus preserved and handed down.

In the seventeenth century the use of powerful magnifying lenses greatly stimulated the study of minute organisms, including those living in fresh water. Although the compound microscope, using two lenses, was known as early as 1621 at least, the simple microscope, using a single convex lens, was preferred by the early naturalists. In the compound microscope the eyepiece lens magnified the defects of the uncorrected object lens, especially its chromatic aberration, so that the image of the specimen being examined was surrounded by colour fringes that made fine detail impossible to see, a disadvantage not so obvious with a simple lens; remarkably fine work was carried out with them.

Antoni van Leeuwenhoek (1632–1723) of Delft, in his leisure time, made and used simple microscopes of various powers consisting of two metal plates, between which was mounted a very small single convex lens. One, still in existence in the Utrecht University Museum, magnified $\times 275$, but so much light was lost with such powerful lenses that only transparent objects could be effectively examined. A drop of water was an obvious choice for study, and thus it came about that Leeuwenhoek discovered Protozoa in 1674 and gave the first description of these simple organisms in a letter written in 1675. In 1683 he gave the first account ever written of bacteria and in 1687, after macerating bruised pepper in rain water, he found rotifers. He described the fixed rotifers *Limnias* and *Floscularia* in a paper published by the Royal Society (of London) in 1705 and 1713. Not only was he the first to discover and describe *Hydra*, but

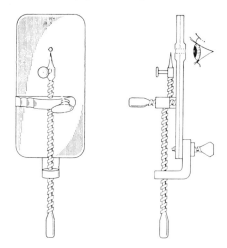

Fig. 1 Leeuwenhoek microscope, *c.* 1673. Front view and view from the side

he also noticed the protozoan parasite now known as *Trichodina* running about the polyp.

Microscopical plants also claimed his attention. He found *Volvox* in 1700, in a ditch, and despite its ability to move about freely in the water, he decided that it was more likely to be a plant than an animal. A paper by him, published by the Royal Society in 1702, contains the earliest figure of a diatom (*Tabellaria*), the cells of which he had thought at first to be salt crystals but later correctly identified as plants.

In Britain, Robert Hooke (1635–1703) using a compound microscope 'with glasses of an English make' published excellent descriptions of the larva and pupa of 'the water-insect' (*Culex*) and the adult 'tufted or brush-horned gnat' (*Chironomus*).

Abraham Trembley (1710–84), a native of Geneva, carried out some experiments in The Hague using only a single lens held in position by an adjustable arm. In 1744, he published most detailed observations as the result of his work on three species of *Hydra*, describing their structure, movements, feeding and digestion, reproduction and regeneration. He also carried out studies on Protozoa (*Stentor*, *Epistylis* and *Carchesium*), the fixed rotifer *Floscularia ringens*, the moss animal *Lophopus crystallinus* and the annelid worm *Stylaria*.

Notable studies of larger animals, especially aquatic insects, were also made in the seventeenth and early eighteenth centuries. Jan Jacoby Swammerdam (1637–80), born in Amsterdam and the son of an apothecary, worked out the life-histories of several aquatic insects. An account of the mayfly *Palingenia longicauda* was published in his lifetime (1675), although his main work, *Biblia Naturae*, appeared posthumously (1737–38).

Johann Leonhard Frisch (1666–1743) of Berlin, who wrote the first important treatise on insects, made detailed studies of the great silver beetle, *Hydrophilus piceus*, in captivity and described the air-space into which the respiratory spiracles open.

René-Antoine Ferchault de Réamur (1683–1757) of France studied many insects including caddis-flies, mayflies and dragonflies. He gave lucid descriptions of these insects, illustrated by superb engravings, in the six quarto volumes of his unfinished *Histoire des Insectes* published between 1734 and 1742.

August Johann Roesel von Rosenhof, known as Roesel (1705–59), expressed the pleasure these early naturalists found in their investigations in the title of his book, *Insekten Belustigung*, which might be translated 'Insect Delights'. His work, published in four quarto volumes between 1746 and 1761, contained observations and illustrations on the water beetles *Dytiscus marginalis*, *Cybister Acilius* and *Hydrocharis*. A supplementary chapter in Volume III gave details of the protozoans *Stentor*, *Vorticella*, *Carchesium*, and the first description and engraving of *Amoeba*, the moss animals *Plumatella* and *Cristatella* and the annelid worms *Nais* and *Stylaria*.

By the early part of the nineteenth century the optical efficiency of the compound microscope had been greatly improved. This was notably through the development of achromatic lenses using a combination of flint glass and crown glass, which corrected the chromatic aberration of lenses made solely of crown glass. The microscope became the invaluable research tool that it has remained ever since.

About this time the discovery of plankton (Gr. = wandering, roaming), the vast assemblage of free-floating and drifting plants and animals, mostly microscopic, in both sea and fresh water, made the microscope of supreme importance and gave a fresh impetus to the study of aquatic life. Although the term plankton was not proposed until 1887 by Victor Hensen, the organisms which it comprised had been known earlier. Johannes Müller, about 1845, designed a tow-net and at a course for zoologists at Heligoland, brought to their notice the wealth of new life to study by drawing the net through the waters of the North Sea. Later Peter Erasmus Müller discovered planktonic microscopic Crustacea in the clear waters of Swiss lakes. It was in the same country that F. A. Forel (1841–

1912), a professor in the University of Lausanne, laid the foundations of the systematic study of the life of lakes, and all its determining factors, by his investigations in Lake Geneva (Lac Léman) between 1868 and 1909. It was he who gave the name limnology (Gr. *limne*, marsh or pond; *logos*, discourse) to the comprehensive study of lakes, but the term has since been applied to the investigation of inland waters generally.

Forel's example gave a further impetus to freshwater investigations and research stations were set up in several European countries and in North America. Britain lagged behind in this field, marine studies taking precedence. It was not until 1931 that an institution comparable to those in other countries was set up (see p. 258).

Limnology is far more than a study of freshwater plants and animals. It comprises all phenomena pertaining to the freshwater environment as a whole. It is, therefore, a composite science taking in its scope not only zoology, botany and ecology, but also chemistry, physics, geology, meteorology and other disciplines in so far as they relate to conditions in inland waters. Its unifying theme is the biological productivity of these waters and the factors which influence it. Some mention of these subjects will be discussed in the following pages.

Limnology has not developed so rapidly as its sister science, oceanography—the study of life in the sea and the factors which influence it. The greater variety and abundance of marine forms of life, compared with those of fresh water, have had a greater attraction for biologists, and governments and institutions have been more lavish in their allocation of funds for marine research than for that concerned with inland waters which, mistakenly as it now appears, seemed to offer little in the way of direct benefit to mankind. The development of oceanography has, however, not been without advantage to limnology, since the increase in general biological knowledge which has been obtained, and developments in the methods of study, have been of assistance to those now engaged on research into freshwater problems.

The oceans, however, are beyond the reach of most of us, and the local pond or stream, both so close at hand, and not requiring complicated equipment to investigate, will provide endless interest and unlimited scope. The very compactness of a pond, in particular, offers a most convenient field of study, for within a very limited area is a world in miniature where, largely isolated from contact with the outside world, the interdependence of the organisms and their relations to their environment, although complex, can be studied. Comparatively little serious research has been carried out on smaller bodies of water, and it is certain that anyone who applies himself seriously to the subject can make valuable contributions to our knowledge of freshwater ecology.

The Invasion of Fresh Water

However life began, there seems little doubt that it was in water that the first living substances had their origin, and it is fairly generally assumed that this took place in the sea. In the aeons of time which followed, some forms invaded and successfully colonized dry land—the beginning of the terrestrial flora and fauna—and others stayed in the water. Of these latter a comparative few, perhaps making their way up rivers or into coastal swamps, eventually reached fresh water, and their descendants comprise the *primarily aquatic* plants and animals which make up most of the present-day flora and fauna of ponds, lakes and streams.

But in addition to these true aquatic organisms—those which have always lived in water—there are representatives from widely differing groups of land plants and animals which have returned to the water to live, finding there, perhaps, more abundant food, less danger from drought, or some other favourable condition which has made it worth their while to effect the change. These *secondarily aquatic* organisms include all the higher water-plants, the aquatic insects, mites and spiders, and some of the water-snails—a heterogeneous collection, all the members of which cannot be considered as anything more than aliens, but which in their separate ways have become adapted to live in their new environment.

This difference in origin of these two types of aquatic life explains many anomalies of structure and habit which will be encountered in freshwater studies. Some insects, for instance, are only aquatic during certain stages of their life, and many of them have to breathe atmospheric air just like creatures living on land.

The term *plankton* to describe the free-floating, mainly microscopical organisms in both salt and fresh water has already been described. Other terms applied to aquatic animals to indicate their spatial position in the water include *neuston,* for animals associated with the surface film; *nekton,* for those that swim actively rather than drift passively as do planktonic organisms; *benthos,* for animals on or immediately above the substratum. *Pelagic* animals are members of the nekton that live in open waters away from the shores and the bottom.

Classification

The main groups, or *phyla* (Gr. *phylon,* a race), of animals found in fresh water are given on pages 9–10, but before listing them it will not be out of place to say a few words on the way in which plants and animals are classified, and why it is necessary to adopt unfamiliar names in scientific writings. Only too often the general reader is deterred by these strange Latin or Greek names, but on closer acquaintance they are found to be necessary and of great assistance in learning more about the plants and creatures concerned.

It must be obvious that it is necessary to have some system of classifying the multiplicity of plants and animals that are found on the earth. Modern classification attempts to take into account not only their relationships to other living things, but also the course through which they have passed during their evolution.

It is necessary also to have some system of naming organisms so that anyone, in whatever part of the world he may be, knows exactly what is referred to by the names. Since Latin and Greek were universally understood by cultured people in all countries, scientific names are based on one or other of these languages, and far from being meaningless are usually very apt descriptions of the creatures or groups to which they refer. It is a great pity that nowadays few people, even those with academic training, take so little trouble to understand what the scientific names stand for.

The first stage in the classification of animals or plants is to group them into kinds, or *species*, of individuals sharing detailed characters which are more or less constant from generation to generation. Groups of species with broadly the same features are included in a *genus* (plural *genera*). Genera are grouped into *families*, families into *orders*, orders into *classes*, and classes into *phyla* (singular *phylum*). Subdivisions and also super-divisions occur throughout the range to indicate smaller or larger groups respectively, e.g. *sub-order*, *super-class*.

In what is called the binominal system of nomenclature (the former term 'binomial' should be confined to mathematics) each species receives two names, e.g. *Dytiscus marginalis*. The first, the generic name, comparable to a person's surname, is always printed with a capital initial letter. The second, printed without an initial capital letter, variously called the trivial name, the specific epithet or the specific name, is comparable to a person's Christian name. To avoid confusion, it is preferable to call the two names together (or the binomen) the specific name. In some groups of organisms three names are sometimes given, the third indicating a geographical race or a variety of the typical form which does not merit full specific rank. The specific name is usually printed in italic type, e.g. *Gammarus pulex*.

To complete the name, when strict scientific accuracy is necessary, the surname of the author who first adequately described the species in print is added at the end. Because of the large number of species that were named by Linnaeus it is traditional to use only his initial, e.g. *Dytiscus marginalis* L. When a species is later transferred to another genus, the name or initial of the original author is placed in brackets.

Scientific names are bestowed strictly in accordance with International Codes, the purpose of which is to ensure that each species of plant or animal has only one name, and that is the first valid one that was legitimately published. Although this sounds simple in theory, it is by no means so in practice, as the first date to be taken as the starting point and the adequacy of description or publication may not be without controversy. However, so far as freshwater organisms are concerned, the tenth edition of *Systema Naturae*, 1758, by Linnaeus is accepted as the starting point. Difficulties arise when, as often happens, the first valid name was published after 1758 in an obscure journal, remained unnoticed for many years, and the animal or plant was later 'discovered' and given another name. When the original source comes to light, the Law of Priority generally makes it necessary to abandon the more recent and probably well-known name in favour of the earlier unfamiliar one. Although this may be necessary in the interest of systematic nomenclature, it is a constant source of difficulty and confusion. Many biologists now feel that the time has come to stabilize the names of existing species, and it is to be hoped that the respective International Commissions on Biological nomenclature will soon find means of doing so. In this book the old names, where they are well known and well established, are given in brackets after the current names with an 'equals' sign, thus: *Floscularia (= Melicerta) ringens*.

The diversity of living organisms in fresh water is not generally realized; with only a few exceptions, all major groups of both plants and animals have representatives in the water. These are listed below; an asterisk * indicates those that are primarily aquatic and a dagger † those that are secondarily aquatic.

PLANTS

*Algae

Cyanophyta: blue-green Algae	Numerous and widespread
Chrysophyta: yellow-green Algae	Including diatoms: numerous and widely distributed
Chlorophyta: green Algae	Numerous and widespread
Rhodophyta: red Algae	Mainly marine but with a few freshwater representatives

*Bacteria

Schizomycetes	Many species of sulphur bacteria, iron bacteria, nitrifying and denitrifying bacteria in fresh water

*Fungi

Phycomycetes	Many species of Chytridiales, Cladiales, Saprolegniales and Pythiaceae are either parasites or saprophytes of freshwater plants or animals, or important in decomposition of terrestrial vegetation, e.g. leaves, in the water

*Bryophyta
Hepaticae: liverworts	A few aquatic genera, e.g. *Riccia*
Musci: mosses	Some aquatic genera, e.g. *Fontinalis*

Pteridophyta
Isoetaceae: quillworts	One genus with a few species, *Isoetes*
Equisetaceae: horsetails	One genus with aquatic species, *Equisetum*
Azollaceae: water-ferns	One genus with two species (introduced to Europe)

†Angiospermae (Flowering plants)
Dicotyledones
Ranunculaceae: crowfoots	Several aquatic species
Nymphaeaceae: water-lilies	All species aquatic
Ceratophyllaceae: hornworts	All aquatic species
Cruciferae: cabbage, cress	A few aquatic species
Lythraceae: loosestrifes	Marginal plants
Haloragaceae: water milfoils	Aquatic species
Callitrichaceae: water starworts	Aquatic species
Umbelliferae: dropworts, etc.	Some aquatic species
Polygonaceae: persicarias, etc.	A few aquatic species
Menyanthaceae: bogbean, etc.	A few aquatic species
Lentibulariaceae: bladderworts	Mostly aquatic species

Monocotyledones
Alismataceae: water plantains	Aquatic species
Butomaceae: flowering rush	Aquatic species
Hydrocharitaceae: frog-bit, etc.	All aquatic species
Potamogetonaceae: pondweeds	Mostly aquatic species
Zannichelliaceae: horned pondweed	Aquatic species
Eriocaulaceae: pipewort	Aquatic species
Najadaceae: naiad family	Aquatic species
Iridaceae: irises	An aquatic species
Araceae: arums	Several aquatic species
Lemnaceae: duckweeds	All aquatic species
Sparganiaceae: bur-reeds	Some aquatic species
Typhaceae: reedmaces	Mostly aquatic
Cyperaceae: sedges	Some aquatic species
Gramineae: grasses	Some aquatic species

ANIMALS

*Protozoa: non-cellular animals
Mastigophora: flagellates	Many species in fresh water
Rhizopoda: amoebas	Many species
Actinopoda: sun animalcules	Mainly a freshwater group
Ciliophora: ciliates	Many species
Sporozoa: parasitic protozoans	Many species

*Porifera: sponges
Demospongiae: flinty sponges	A few species

*Coelenterata: stinging animals
Hydrozoa: hydras	A few species

*Platyhelminthes: flatworms
Turbellaria: planarians	A few species
Trematoda: flukes	A few species parasitic in freshwater hosts
Cestoda: tapeworms	A few species parasitic in freshwater hosts

*Aschelminthes: worm-like animals
Nematoda: roundworms	Many aquatic species
Nematomorpha: hairworms	Some species as parasitic larvae and adults
Rotifera: rotifers	Mainly a freshwater group
Gastrotricha: hairy-backs	Wholly freshwater group

*Acanthocephala: proboscis roundworms
	A few are parasites in freshwater hosts

*Ectoprocta: moss animals
Phylactolaemata	A few freshwater species

Mollusca: molluscs
*†Gastropoda: snails	Many freshwater species
*Bivalvia: mussels	A few freshwater species

*†Annelida: worms
Oligochaeta: worms with few bristles	Many species in fresh water
Hirudinea: leeches	A few species in fresh water

Arthropoda: jointed-limbed animals
*Crustacea: crustaceans	Many species from five sub-classes in fresh water

†Insecta: insects	Many species from eleven orders in fresh water
†*Arachnida: spiders and mites	One species of spider and many species of mite are aquatic

*Tardigrada: water-bears   A few species in fresh water

Chordata:

*Cyclostomata: lampreys	Two species in fresh water
*Pisces: bony fishes	Many freshwater representatives
*Amphibia: amphibians	Species of newts, frogs and toads, aquatic in their early stages
Aves: birds	Many closely associated with fresh water habitats for food, etc.
Mammalia: mammals	A few mammals live in close association with freshwater habitats

2 : THE FRESHWATER ENVIRONMENT

The influence of physical and chemical factors on the lives of plants and animals can be seen more clearly, perhaps, in the freshwater environment than in any other, and the most important of these factors must now be considered.

Chemically pure water would not support life, but its ability to dissolve more substances than any other liquid in nature ensures that natural waters usually contain in solution those substances needed by living organisms.

The sight of fresh water in ponds, lakes, streams or rivers is so familiar that few people pause to reflect that it has all come originally from the atmosphere in the form of rain, hail or snow, and is being replenished continuously from the same source. Comparatively little falls direct into these bodies of water, most finding its way by springs and drainage from surrounding land. During its journey through the atmosphere and the ground, and as it stands in the lake or pond, water dissolves or acquires the solids and gases that enrich it and make it suitable for living things. It may also, of course, gather substances such as poisonous effluents from factories that are detrimental to life. The nutrients that any particular stretch of water has acquired, and their quantities, determined largely by the geology of the catchment area, have a considerable influence on the plants and animals that can thrive there. On different soils or under different conditions, therefore, quite distinct communities of living organisms may be found.

Chemical factors

The chemistry of fresh waters is a very complex subject and any detailed account would be out of place in a book of this nature. A very brief mention of the more important substances that influence life in aquatic habitats is all that can be attempted here, and the reader is referred to more advanced works mentioned in the bibliography on page 268 if he or she wishes to pursue the subject further.

Oxygen With the exception of a few anaerobic organisms (see p. 48), all living plants and animals need oxygen for respiration. Water is, of course, a compound of oxygen and hydrogen, but it is not the O in H_2O that is available for this function but the quantities of free oxygen held in solution, partly by being dissolved from the atmosphere at the surface where air is in contact with water. Although oxygen occupies about one-fifth of a given volume of air, this is about twenty-five times the concentration of the gas in the same volume of water. The quantity dissolved at any time depends on the temperature of the water, the partial pressure of the gas in the atmosphere, and the concentration of dissolved salts in the water. At a normal air pressure of 760 mm the amount of oxygen in pure water, when fully saturated, is about 12·5 mg per litre (or 8·75 ml per litre) at 4° C, but only about 8·9 mg per litre (6·3 ml per litre) on a warm day with a water temperature of 20° C.

Diffusion of oxygen from the surface to lower depths is a slow process, but violent agitation of the surface by, for example, wave action in lakes or a river foaming over a waterfall, causes more oxygen to be dissolved and diffused more quickly. The use of an aerator in an aquarium is based on this principle; the air is not, as is often thought, being 'pumped' into the water.

The augmenting of the oxygen supply in the water by the photosynthesis of green plants, both microscopical and small, is an important factor, especially in small ponds. In a weedy pond on a bright, sunny day in summer, it is common to see bubbles rising from the plants, and analysis shows that such waters are super-saturated with oxygen. As algae produce oxygen which goes straight into solution, they are even more likely to cause super-saturation. Nevertheless, during the night, when photosynthesis temporarily ceases, the dissolved oxygen in ponds may be almost completely used up by oxygen-consuming processes of which the respiration of the animals, as well as of the plants themselves, are important factors; it must not be forgotten that plants need oxygen for respiration just as animals do, and this process goes on day and night.

In waters which contain large quantities of decaying animal and plant matter, or where there is organic pollution, there may also be a shortage of oxygen, for the gas is used up in the process of decomposition by the respiration of micro-organisms. The deficiency may be such as to render the water unsuitable for animal life, except for a few species which have adaptations for existing in conditions of oxygen depletion. The black, evil-smelling ponds sometimes encountered are typical of such conditions. Often they are surrounded by overhanging trees, the dead leaves from which, accumulating for many years on the bottom of the pond, have provided such a store of decomposing matter that oxygen is almost absent from the water. The pond-hunter can safely ignore such ponds unless he is in search of certain aquatic worms or the larvae of one or two insects, such as the rat-tailed maggot (see p. 209), all of which are well adapted for living in the black mud at the bottom of these unwholesome waters.

Most of the simpler aquatic creatures absorb oxygen through the whole of their external membranes, or 'skin', whereas others have special areas set apart, or organs such as gills where this can take place. The most interesting and ingenious methods are, however, adopted by those creatures, such as the insects, which are not primarily aquatic, but have invaded the water at a late stage in their evolution; these will be dealt with more fully later in the book.

Nitrogen Nitrogen is important in nature for the building of proteins which, together with carbohydrates and fats, are the main constituents of protoplasm—the living substance. The gas occupies about four-fifths by volume of atmospheric air compared with approximately one-fifth of oxygen, but under similar conditions of temperature and pressure the solubility of nitrogen is only about half that of oxygen. In most freshwater habitats elementary nitrogen is usually at saturation point, and a range of between 12 and 20 ml per litre has been recorded in lakes. Although there are organisms, such as some bacteria and blue-green algae, that can 'fix' free nitrogen, most freshwater plants and animals derive their requirements of the element from nitrogenous compounds. These include nitrates, nitrites and ammonia, which are either washed into the water from the land, or that result from the decomposition of organic matter in the habitat itself. The role played by bacteria in breaking down the complex proteins in dead animals and plants, first to ammonia, then to nitrites, and finally to nitrates, thus making an assimilable form of nitrogen available once more, is described briefly in Chapter 4.

Carbon Dioxide Although this gas is present in the atmosphere in only minute quantities (about 0·03% by volume compared with about 21% of oxygen and 79% of nitrogen), its role in the living world is all important. In the process of photosynthesis green plants, using the energy of sunlight, combine carbon dioxide with water to produce major constituents of all living protoplasm: carbon, hydrogen and oxygen.

Carbon dioxide is very soluble in pure water and, at 4° C, the solubility is about 0·94 mg per litre and, at 20° C, 0·56 mg per litre at normal pressure. Comparatively little of the gas is absorbed from the atmosphere at the surface of standing water but, dissolved in raindrops, it reaches aquatic habitats either directly, or as ground water percolating through the soil or over the surface. The ground water collects further supplies of carbon dioxide from the respiration of plant roots, and also of micro-organisms, causing the decomposition of organic matter with which it comes into contact. Further quantities are obtained from decomposition in or on the mud of the aquatic habitat itself. Finally, the respiration of larger plants and animals in the water releases the gas into the environment. Carbon dioxide combined chemically with water forms carbonic acid. When this weak acid comes into contact with rocks or soils containing carbonate, of calcium or magnesium for example, the carbonates are converted to loosely combined bicarbonates.

The gas, therefore, can exist in natural waters and muds in three forms: free, in solution and as carbonic acid; in loose association as soluble bicarbonates; and locked in more stable compounds as carbonates, which sink to the bottom or are sometimes precipitated on solid objects such as stones or plants (see p. 14). The carbonates serve to 'buffer' the water against sudden changes of hydrogen-ion concentration or pH (see p. 263). If the amount of carbon dioxide in the water increases, as for example by respiration of plants and animals, some of it will combine with the carbonates to form bicarbonates and raise the pH. If carbon dioxide is withdrawn from the water by the photosynthesis of plants, some of the bicarbonate will be converted to carbonate, thus replenishing the supply of carbon dioxide and lowering the pH. Nevertheless, in small, weedy ponds there can be wide fluctuations in pH daily or seasonally.

Calcium and Magnesium   It is the amounts of bicarbonate and carbonate of calcium and magnesium in natural waters that determine their 'hardness'. Calcium is usually the more abundant ion and is derived from lime-bearing rocks and soils which include not only the obvious limestones and chalks, but also marls and even basalts. Waters with less than 7 mg per litre of calcium are regarded as 'soft', those with quantities of between 7 and 24 mg per litre as intermediate and those with over 24 mg per litre as 'hard'.

It has long been known by gardeners that some plants need lime (calcium carbonate), whereas others do better without it. It took longer to realize that animals, too, have similar preferences and can also be described as calciphile (L. *calus*, lime, chalk; Gr. *philos*, fond of), or calcifuge (L. *fugio*, to flee).

Some snails and crustaceans, for instance, are found only in very hard waters because the shells of snails and external skeletons of Crustacea need a good deal of calcium. Freshwater sponges and moss animals are among animals that do not tolerate high degrees of hardness. In general, however, hard waters are more biologically productive than the soft, calcium-deficient waters of moorland bogs and mountain tarns. The chemical stability of waters rich in bicarbonates and carbonates, as mentioned above, may be a not unimportant reason.

Sometimes, in hard waters, a chalky coating is found on plants or other objects. This is due to the photosynthetic activities of plants by which carbon dioxide has been withdrawn from soluble bicarbonates, as explained earlier, and carbonate precipitated. Stoneworts (p. 64) derive their name from the phenomenon, but *Elodea*, *Stratiotes* (p. 43) and species of *Potamogeton* (p. 39) are other plants in which this has been observed. The green algae *Cladophora* (p. 60), normally a branching, filamentous plant, under some circumstances forms into balls, with a diameter up to several centimetres, around a core of chalky material.

Silicon   The element silicon does not occur in nature in a free state, but the many forms of silicon dioxide, or silica, are among the commonest constituents of rocks and soils. Quartz is perhaps the best known, with sea sand as its most familiar form. The forms of silicon that are of interest to the limnologist are the soluble silicates which reach lakes by inflowing waters. They are important because the frustules or 'shells' of diatoms, the microscopic plants which are the first links in the food-chain of lakes, are almost entirely made of silica, and their growth is dependent on the supplies available in the water. As the period of maximum diatom production proceeds (p. 23), the amount of silica decreases until, when it reaches a quantity of about 0·5 mg per litre, the numbers of such diatoms as *Asterionella*, *Tabellaria* and *Melosira* (Chapter 4) fall steeply. The shells of dead diatoms, including those that have been eaten by small animals and passed through their digestive system, drop to the bottom of the lake where some of the silicate slowly dissolves and comes into circulation again. The shells are very resistant, and their ability to persist in recognizable form for thousands of years has enabled the history of lakes, and of their catchment areas, to be deduced from examination of mud cores. The needle-like spicules which form the skeletons of freshwater sponges are also made of silica, and their shapes and forms are used as a basis for classifying the species (Chapter 6).

Other mineral substances to be found in small, but important, amounts in natural waters are phosphates (of calcium, iron and magnesium), chlorides (of sodium, potassium, calcium and magnesium) and sulphates (of potassium, sodium, magnesium and calcium).

Organic substances, mainly in solution, but also in particulate form, are present in the fresh waters of lakes in greater amounts than in the sea, and include proteins in colloidal solution and several amino acids. They are an important source of food for some animals, including Protozoa. Finally, trace elements (so called because only minute traces of them occur) such as manganese, boron, molybdenum, vanadium, cobalt, copper, nickel, lead and titanium have been found and may be important biologically; the first six are necessary for growing some algae in culture.

Seasonal fluctuations occur in the amounts of many of the chemical substances in the water. Nitrates, resulting from the autumnal decay of land vegetation, are carried into the water with the heavy rains of autumn and winter, as are phosphates and silicates. By summer their quantities have been much reduced by the spring and early summer outbursts of planktonic algae. The effects of thermal stratification in lakes on the supplies of nutrients available to plants are discussed on pages 21 to 23.

Mineral salts, and particularly the chlorides, are present in fresh water in much smaller quantities than in sea water. The body fluids of freshwater creatures, however, usually contain larger amounts of the salts than does the surrounding water, but generally well below those in marine animals.

The consequence of having two solutions of different salinity separated by a membrane which allows both salts and water to pass is that water will diffuse into the strong solution from the weak, and salts will diffuse out of the strong solution to the weaker until equilibrium is reached and the two solutions are of equal salinity. Should, however, the membrane be permeable to water only, and not to salt, then water will pass into the stronger solution from the weaker only. The external membrane of most aquatic creatures is of this latter type, and as they have a high salt content within them and a low concentration in the outside water, they must have a means either of ridding themselves of the water which diffuses into the body or of preventing the ingress of the water.

The higher creatures have kidneys, which excrete the surplus water. Simple single-celled creatures, such as *Amoeba* and *Paramecium*, usually have large expandable bubbles or spaces in the midst of the protoplasm, called contractile vacuoles, in which excess water collects and is then passed out of the creature. Some organisms have slimy, gelatinous mucus or coverings which protect them against the entry of the water.

A steady supply of salts is essential to living creatures, and many structures, the purpose of which had been in doubt, are now known to enable the animals to maintain osmotic pressure and the balance of salts between the blood and the surrounding water. The leaf-like structures on the tail end of gnat larvae and the outgrowths at the tail end of the 'bloodworm' larva of *Chironomus* are such organs, and there is little doubt that some of the so-called 'gills' of aquatic creatures are concerned not with respiration, but with salt assimilation.

Physical Factors

The physical characteristics of fresh water are no less important to the organisms living in it than are the chemical.

Density Water is over 700 times heavier than air, and its density nearly corresponds to that of protoplasm. The more minute planktonic plants and animals can float effortlessly, although in summer others develop extensions of their structures, which, it has been suggested, may compensate for the reduction of density, and therefore of buoyancy, that higher temperatures bring about.

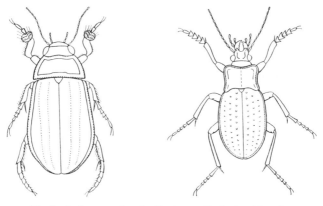

Fig. 2　Left: water beetle, *Dytiscus. Right:* land beetle, *Carabus*

Larger animals in the water are relieved of the necessity of supporting their own weight; their limbs can be used solely for swimming. To offset the resistance offered by the water, animals that have to move quickly have streamlined bodies. One has only to compare the two beetles illustrated above to know at once which is adapted for swift movement through a denser medium.

The buoyancy of water enables soft-bodied creatures such as hydra and the moss animals to extend themselves in a way that would be impossible in air. Water plants do not need the strong supporting tissue that their land relatives must have to remain upright.

A property of water which has been little studied in its relation to the life of freshwater creatures is its excellence as a conductor of sound and vibrations. Some insects, such as corixid bugs and the screech beetle, *Hygrobia*, have stridulating organs by means of which they can make sounds in a similar manner to a grasshopper, and these are believed to be used in attracting the opposite sex.

Surface Tension The surface film on water—the layer which, although not differing in chemical composition from the rest of the water, is yet in a peculiar physical state and acts rather like an elastic skin—is of some importance to small creatures. The time-honoured experiment of floating a needle on the top of a tumbler of water demonstrates the property of the film in supporting solid

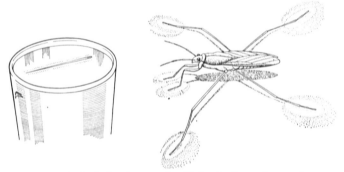

Fig. 3 Left: floating needle. *Right:* the legs of a pond skater, *Gerris*, merely depress the surface film

objects above it, and explains how some of the water bugs can walk or glide about on the surface as confidently as though they were on a sheet of glass. The film will also support small objects from its underside. Some creatures move about underneath the film, as much at home as a fly walking on the ceiling, and others, such as gnat larvae, hang from it down into the water. The tension which causes the film also prevents water from penetrating into minute openings or finding its way into hairy structures, a property vital to the breathing of some creatures. The surface tension can be upset by adding oil or detergents to the water, and this method is, as we shall see later, employed by man in the control of mosquito larvae.

Transparency The transparency of water is important in enabling green plants, both the rooted higher plants and the planktonic algae, to obtain sufficient light to carry on the process of photosynthesis. In clear lakes on the Continent adequate sunlight was found to penetrate to depths of over twenty metres, although in most waters the turbidity caused by the presence of living and non-living particles in suspension reduces the transparency considerably, and the nine metres recorded for Wastwater in the English Lake District is a high figure. Light, too, has a direct influence on the movements of some small aquatic animals. This can be observed by keeping a number of small crustaceans, such as *Bosmina*, in a glass jar which has one side shielded from the light. They will be found to congregate on the lighter side of the jar. Too strong a light, however, will have the reverse effect, and will drive them to the less brightly lighted areas of the water.

Temperature Water is a poor conductor of heat. Because of its high specific heat, it takes a longer time to warm up than does air, and is slower in cooling, so that fluctuations in temperature are neither so great nor so violent in water as in air, a matter of some importance to creatures entirely dependent on their surroundings for their body temperature.

Everyone must have noticed that when water in a pond or lake freezes there is usually only a thin layer of ice on top. Very rarely in this country does the water freeze to the bottom. This is due to a peculiar property of fresh water—that on cooling it contracts and becomes denser only until it reaches a temperature of 4° C (about 39·2° F). If cooled further, it begins to expand again and becomes lighter. Hence on the approach of freezing conditions, the colder waters from the surface sink at first and warmer water from below rises to take their place. This continues until the temperature of all the water is at 4° C, when further cooling results in the colder water remaining at the surface, where it freezes. The deeper water, however, remains at the temperature of 4° C and the layer of ice now on top of the water itself acts as a blanket and slows down further cooling. Even in the severest weather, therefore, there is usually an unfrozen area of water in which the creatures can live.

An important indirect effect of temperature on freshwater organisms is the varying quantities of oxygen that water can hold in solution as it becomes warmer or colder; this aspect was discussed on page 12.

Movement The movement of the water, whether as the flow of a river or stream, or as wave-action beating on the shore of a lake, affects the plant and animal life, mainly through its effect on the substratum, the material on the bottom over which the water is passing. With violent movement, small particles will be carried away and only stones left. Where the flow is gentler, sand particles or even mud may remain to enable higher plants to take root. The clear, stony beds of mountain streams, the gravelly bottoms of the lower reaches, the silted bends of sluggish rivers and the muds of various types of ponds, all have their characteristic flora and fauna.

The temperature of running waters remains more constant, even if lower, than still waters and, as we have seen (p. 12), cool water contains more oxygen than warm water. The disturbance to the surface in moving water also causes more air to be dissolved so that generally, unpolluted rivers or streams are better oxygenated than stagnant waters. The mayfly and stonefly nymphs, which habitually live in fast streams, soon die if kept in water warmer than that to which they are accustomed. Animals living in conditions where water movement is violent are adapted for clinging or otherwise holding on to the stones or rocks, and are sometimes flattened in form to minimize the effect of the current.

Impermanence   It is a matter of common observation that not all ponds, pools and streams are permanent. For various reasons some dry up in all but the wettest summers, and one might expect that a repeated catastrophic occurrence of this nature would preclude the existence of aquatic creatures in such waters. This, however, is not the case. Many of the simpler organisms such as tardigrades, worms, Protozoa and phyllopods (fairy shrimps) have means of tiding over such periods. So well adapted are the latter animals to temporary conditions that they are rarely found in permanent waters, but only in pools which cannot fail to dry up.

The permanence of even large and established bodies of water is illusory when seen in the context of geological time. All freshwater habitats have built-in features of self-destruction, even if the processes are so slow that they are hardly apparent in the span of a human life. In ponds and small lakes the growth of marginal plants, and the accumulation of their decaying vegetation, gradually raises the level of the ground at the edges and confines the water to a smaller and ever-decreasing area, until eventually a swamp or marsh is formed. In large lakes, wave action slowly grinds down the rocky margins, and the silt which is deposited gradually raises the level of the bottom, while streams carry more and more silt into them. Streams and rivers are ever widening their courses, reducing the speed of their current and ultimately silting up. All these changes in the physical conditions bring about changes in the animal and plant life, gradual though they are.

Types of Freshwater Habitats

Having examined, if only briefly, some of the factors affecting life in fresh water, consideration can now be given to the various types of environments in which aquatic life is found, and see how these factors influence the organisms found there.

Lakes The word 'lake' is used somewhat loosely in common parlance to mean almost any largish stretch of still water, such as a boating 'lake' in a public park or an ornamental pond in the private grounds of a big house. References to lakes in books on freshwater biology, however, usually mean the large, naturally occurring body of water found in the Lake District and elsewhere, in which there is a wide expanse of open water exposed to the full force of the winds and a depth too great in the middle for higher plants to grow. Owing to these features, such lakes have certain important characteristics not found in smaller bodies of water. They have come into existence in several different ways. Most of the lakes of the English Lake District, the lochs of Scotland and the llyns of the Welsh mountains were formed by glacial action. The glaciers gouged out a deep rock basin which, when the ice retreated at the onset of warmer conditions, filled with water and was sometimes dammed by a plug of moraines deposited at the front of the original glacier. In less mountainous country, a melting ice-sheet deposited on a plain left a shallow depression which became a lake, such as Loch Leven in Scotland, and on a larger scale the extensive lakes of Wisconsin, U.S.A., and those on the plains around the Baltic were created. The meres of the Cheshire plain were formed in a similar way, in clay hollows, although land subsidence through the removal of natural salt deposits has also played a part in providing freshwater habitats in that district. The so-called tectonic lakes, an example of which is Loch Ness in Scotland, were caused by extensive faulting of the Earth's crust to form a long, narrow and deep basin with rocky sides. Lough Neagh, in Northern Ireland, the largest stretch of fresh water in the British Isles, resulted from the sagging of a huge mass of volcanic lava and the subsequent filling of the hollow with water. The shallow lakes of the Broads, in East Anglia, were formed by the natural flooding of medieval peat-diggings.

Lakes have received more detailed study than other bodies of fresh water, and research has been directed mainly to the free-floating planktonic plants and animals in them, and to the physical and chemical factors which influence their productivity. As a result of these investigations, it has become possible to arrange lakes in a series based on their primary productivity—their annual production of algal material.

At one end of the series is the highly productive lake, typically in an area of softer rocks, as shown by the gentler contours and the presence of good agricultural land. The drainage entering the lake is rich in dissolved nutrients, with a considerable amount of matter in suspension. In the course of time the bottom has become covered with a deep layer of mud, and conditions are favourable for the abundant growth of plant life. In the English Lake District, Windermere and Esthwaite are examples of this type of lake. (Plate 1)

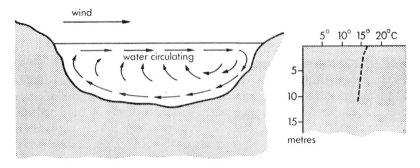

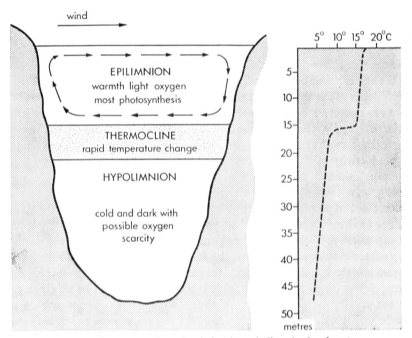

Fig. 4 Above: complete circulation in a shallow body of water.
Below: stratification in a deep lake

At the other end of the series is the lake in surroundings of hard, often volcanic rocks, with land of limited value for agriculture and the drainage from
which yields little in nutrients for plant growth. The lake water is clear and soft,
and the bottom stony or rocky, with little aquatic vegetation. Ennerdale Water,
Wastwater and Buttermere in the Lake District, and Llyn Ogwen in North
Wales, are examples of this type of unproductive lake. (Plate 1) In between these

two extremes are lakes varying greatly in their productivity owing to local differences in geology, climate and human activities.

A factor of the greatest biological importance in large, deep lakes is the existence in summer of layers of water at different temperatures. Let us consider for a moment how this can happen. We have seen how water is at its heaviest when it is at a temperature of 4° C, and in lakes in this country most of the water reaches this temperature during the winter and mixes freely by wind action. In late spring, the returning strength of the sun's rays warms the surface water, but the absorbing power of the water prevents light and heat penetrating into the lower depths, so that two main layers are formed, the upper warmer and lighter waters, and the colder and heavier layer nearer the bottom. Winds circulate the water to a certain extent, but their effect is limited to the upper level of the lake, and a somewhat ill-defined intermediate layer of water, just outside the influence of wind action, comes into existence, throughout the whole depth of which the fall in temperature is very rapid. This layer is called the thermocline (Gr. *thermos*, heat; *klino*, to turn aside, bend), or mesolimnion (Gr. *mesos*, middle), while the water above is called the epilimnion (Gr. *epi*, upon; *limne*, pond, lake) and the colder water below is called the hypolimnion (Gr. *hypo*, under) (Fig. 4).

The depth of the thermocline varies with the type of lake, but it is usually at about ten metres in British lakes. Its importance is due to the fact that it is a barrier effectively preventing the free mixing of waters and their dissolved chemical substances and gases between the upper and lower layers, and limiting the productivity of the lake to the relatively smaller upper layer. The hypolimnion is cold and dark, and little or no plant life can grow there. The epilimnion, on the other hand, is well lighted, warm and well oxygenated, and in a rich 'evolved' type of lake, a wealth of minute plants is enabled to thrive and produce food on which an abundant population of microscopical animals such as crustaceans, rotifers and Protozoa, can thrive and multiply. On the death of the planktonic animals and plants, the insoluble portions of their remains fall like rain into the hypolimnion, where the respiration of the bacteria causing their decomposition uses up the oxygen that is there. As there can be no circulation between the layers, the products of their decomposition (see p. 47) do not become available during the summer to enrich the epilimnion, and as the season advances the depletion of the nutrients there may limit the number of organisms it can support. A lake where such conditions exist is called a eutrophic lake (Gr. *eu*, good; *trophe*, food).

On the other hand, where there is a scarcity of the necessary substances which are needed for plant growth as occurs, for instance, in our second type of lake, namely that on hard, insoluble rocks, a sparse plankton is produced in the upper layers of the water, and the decomposition of its remains on the bottom does not use up all the oxygen during the season. Such lakes are called oligotrophic (Gr. *oligos*, few, small) lakes.

A third type of lake—the so-called dystrophic lake (Gr. *dys*, difficult, bad)—is found in ill-drained mountainous areas and characterized by an abundant growth of sphagnum moss which, in its partial decomposition, produces acips and tannins. These provide an environment that few larger plants or animals find acceptable and such habitats, which include some of the Scottish lochs, are rather barren.

Human activities are capable of changing the productivity of natural waters much more rapidly than natural processes. Some of the manures and fertilizers applied to agricultural land inevitably find their way into streams and lakes, and enrich the water. On the other hand, toxic chemicals used in pest control may poison it. Increasing populations in areas of high scenic beauty similarly enrich the natural waters through sewage effluents, treated or untreated. Industrial wastes, in some parts of the world, have changed the character of natural waters in striking and often disastrous ways (p. 256).

In autumn, with the cooling of the surface of lakes, the whole body of water becomes of the same density, and therefore capable of being mixed freely by wind action. In this overturn the substances which have been accumulating in the hypolimnion are brought into circulation once more. At the same time, the greater abundance of oxygen now in the bottom waters aids the more rapid decomposition of any plant and animal remains there, which further adds to the reserve of nutrients in the lake.

The large quantities of dissolved and suspended inorganic and organic matter washed into the water by autumn and winter floods, together with the products of decomposition brought into circulation at the overturn, provide conditions favourable for the abundant growth of minute plant life as soon as light and temperature increase in spring.

The algae, and particularly the free-floating diatoms, are the first organisms to become plentiful, multiplying rapidly with the first spring sunshine. This burst of plant life provides an abundance of food for the small planktonic animals, which also increase and whose numbers remain high until autumn. The blue-green algae usually become plentiful in summer, and often particularly so in autumn, by which time there are in the water large quantities of the organic substances they appear to need.

The seasonal variations in the different groups of planktonic organisms are often sudden and spectacular, sometimes causing the water to turn green, to develop a scum on the surface or, worse still, to develop an unpleasant smell.

Such, then, are the main characteristics of lake biology, a study, not only of absorbing interest in itself, but one which is benefiting mankind directly in no uncertain manner. For the huge reservoirs which are needed to store the water supplies of urban populations show many of the characteristics of lakes, and the knowledge gained in lake studies is being applied daily to the maintenance of a pure supply of water.

c

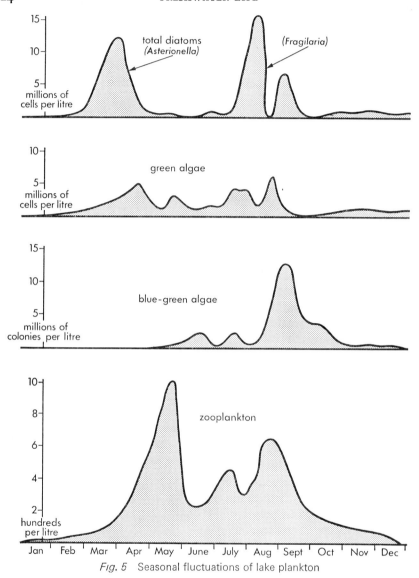

Fig. 5   Seasonal fluctuations of lake plankton

Ponds There is no satisfactory definition of a pond, for the term covers such a wide variety of freshwater habitats. The outstanding characteristics of ponds, from the biological point of view, are that they are *small* bodies of *shallow, stagnant* water, usually well supplied with aquatic plants. Many are the result of man's activities in the past, in providing drink for his animals, or in draining his fields. When cleaned out, these old ponds sometimes reveal a well or spring that provided the water supply, often protected by a stone slab or stone-built spring-head. Other farm ponds are supplied by field drains. Some villages had communal fish ponds, run on similar lines to monastic fish ponds, or the stew ponds on private estates, to provide coarse fish for eating.

The so-called dewponds which were formerly abundant on sheep-grazing downland in Berkshire, Hampshire, Sussex, Wiltshire and other chalk areas, were artificially made to provide drinking water for sheep and cattle on permeable land where water would not collect naturally. They were usually saucer shaped and sited in natural hollows of the down. Various techniques were used by the family firms who made them, but generally there was an impermeable bottom layer of clay, into which flints or other stones were rammed to prevent the hooves of animals penetrating it. On top of this was placed another layer of clay. Sometimes straw was incorporated, possibly to prevent the clay from drying too quickly and thus cracking, or from being affected by frost during construction. The diameter of ponds varied from about seven to twenty metres, and the depth was about two metres. The ponds were carefully sited and designed to receive the maximum amount of rain and surface water and this, with possibly some condensation from low cloud or mist, was the main supply. Dew cannot have provided any significant quantity of water, for at night the water temperature would be higher than that of the surroundings; during short summer nights the surface would very rarely reach dew-point for condensation to occur. The belief that dewponds were of great antiquity is not supported by available evidence. Some dated from the seventeenth century, but many were made in the nineteenth century. One firm of pond-makers was known to be operating as late as the 1930s.

Similar to ponds, in their biological characteristics, are quieter parts of canals with little traffic to disturb their margins. These often have a very rich fauna.

The stillness of a pond is proverbial and in these small bodies of water, sheltered as they often are by trees or high banks, water movements are slight. The abundance of submerged water plants also restricts the free circulation of water. Temperatures tend to fluctuate greatly even over the period of twenty-four hours, and it is an easily observable fact that, in winter, ponds form a coating of ice long before one forms on a lake. Sunlight can penetrate to the bottom of ponds so that rooted plants are able to grow in even the deepest parts, and their presence is an important influence on the abundance of animal life. In some weedy ponds oxygen may be almost exhausted during the night, when the plants are respiring but not photosynthesizing, and only the resumption of photosynthesis, when light returns, rescues the animals from suffocation.

The diversity and abundance of both plants and animals in a good pond are very striking. Four main zones are apparent: the surface film, the vegetation zone, the open water and the substratum, or bottom mud. The principal animals found in each zone are shown in Fig. 6, and a typical food-web in Fig. 7.

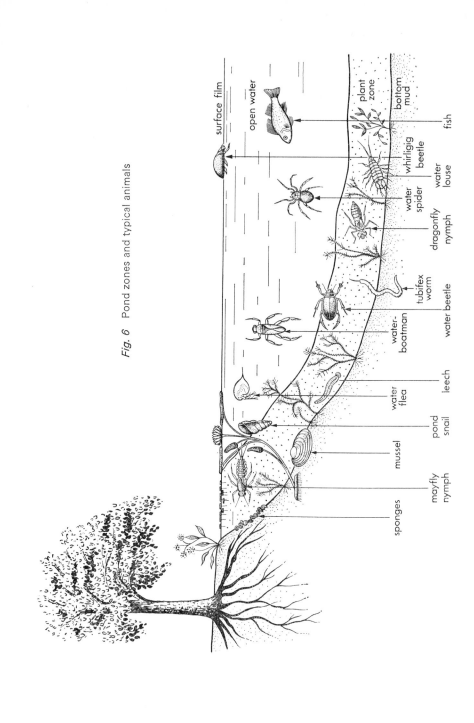

Fig. 6 Pond zones and typical animals

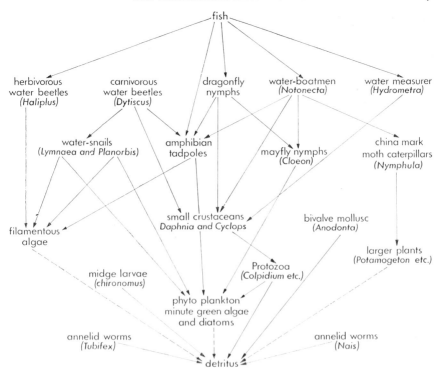

Fig. 7 Food-web of a pond. The dotted lines indicate
breaking-down processes

Planktonic plants and animals are of less importance in ponds than in lakes. Animals that either live among the plants, or on the bottom, are the most characteristic forms, especially insects and their larvae or nymphs. Species which are plentiful in one year may have completely disappeared the next, but may reappear in their original abundance three, four or more years afterwards. No entirely satisfactory explanation of these puzzling changes in populations has been forthcoming.

The study of ponds and pond conditions has been greatly neglected. To professional biologists lakes have always held the greatest attraction, and yet ponds offer a field of research equally challenging. The complexity of pond conditions, and especially the rapidity with which these can fluctuate, present many problems for detailed ecological study involving continuous recording and monitoring. In more restricted fields, such as the elucidation of life-histories, ponds are particularly suited to the activities of the amateur naturalist and, if carried out systematically and carefully, these may well result in valuable contributions to science.

Unfortunately, natural ponds of all kinds are becoming scarcer every year.

Inevitably, some disappear as more land is taken for building and other developments. Piped water supplies in even remote fields have made both farm and
village ponds unnecessary for their original function, as watering-places for
cattle and horses. Campaigns to produce cleaner milk and to reduce cattle
diseases, in particular fascioliasis (see p. 97) and Johne's disease, which is
spread when cattle drink from waters tainted with the dung of affected animals,
have given official discouragement to the use of ponds for this purpose. When
the marginal vegetation of a pond is no longer trampled down regularly by
animals' hooves, and it is no longer worth the farmer's trouble to keep his ponds
open, the natural succession of the hydrosere takes over (p. 45) and ponds
change first to marshy areas, and then into dry land, unless they are filled in
even more quickly by the farmer. At the present time, a government grant is
available for filling in ponds when they impede cultivation and occupy space on
which crops could be grown. A possible solution to the problem of disappearing
ponds is mentioned on page 257.

Streams and Rivers   Running water provides an environment quite different
from that in lakes and ponds. The flow, always in one direction, produces many
effects which determine the kinds of plants and animals that can live in such a
habitat. Just as there are many types of pond and several kinds of lake, so there
is great diversity in running-water environments, and the rushing mountain
torrent, the babbling brook, the shallow stream running through the meadows,
and the wide and sluggish river, all have their characteristic fauna and flora.

In practice the term 'river' is applied to natural watercourses over five metres
wide; anything below that width is a 'stream'.

Several attempts have been made to classify the various reaches of a river into
biological zones. Some made use of the species of fish that were encountered in
different stretches of water, and terms such as 'minnow reach' and 'trout beck'
were employed. Another classification was based on the plant communities that
occur along the watercourse. There is no typical river, however; each one is
unique, so that such classifications have little biological value. Nevertheless, it
may be helpful in showing the range of habitats that rivers can provide, to consider a hypothetical river from its source in the hills to its entry into the sea, and
to review the characteristic plants and invertebrate animals that may be found
in the different reaches. The river will, perhaps, originate from a spring or a
mountain tarn, and a number of almost imperceptible trickles of water may need
to unite before we can even detect that here is a stream in the making. The water
will be shallow, flowing now over soft earth, now over stones. Plants will be
restricted to clumps of moss, a coating of slimy algae on the stones, and perhaps
here and there a growth of liverworts. Such an unpromising stretch would be
passed over as lifeless by most people, but these headstreams have a fauna all
their own. Hiding in the moss tufts is a variety of small creatures that can strain

from the current food in the form of small suspended particles, or can feed on the moss itself—some Protozoa, a few rotifers and insects. Crawling over the stones and scraping the algae off them will be snails; on lifting some of the larger stones, numbers of caddis cases and perhaps planarian worms will be revealed.

The headstreams meet to form a watercourse, wider, deeper and swifter. The current is such that it carries all before it, and sediment is swept along without a chance of settling on the bottom. The stony bed is even more restricted in its plant life, and free-floating plankton could not exist in this turmoil. It is in such stretches that the most beautifully adapted of all the aquatic creatures are found, the flattened forms with special means of clinging to the stones. Only thus can they escape being swept away. The nymphs of such mayflies as *Ecdyonurus* and of the stoneflies are the most characteristic members of this fauna, which includes also some caddis larvae, among them those that build net-snares rather than cases, a few snails and some leeches. (Plate 3) Many of these stream animals are especially active at night, when they emerge from their daytime hiding places under stones, and are carried by the force of the current downstream. This *drift fauna*, which varies in composition and abundance at different times of the year, is obviously an important part of the food of larger animals in the stream, particularly the fish.

Severe though the conditions are in these hill streams, they have their compensations. In particular, the oxygen supply is plentiful through the constant change and splashing of the water, while the temperature remains almost constant at all times, even if at a low level. Creatures removed from such waters rapidly die in the relatively high and fluctuating temperature of an aquarium, and are soon in obvious distress at the scarcity of oxygen there.

One aspect of stream life that is of interest to the student of freshwater life is that there is less cessation of activities in winter than there is in still waters, and these stretches produce their most abundant animal life then. It is of little use trying to collect stream forms of insects, for instance, in summer, for by this time of year many will have completed their metamorphoses and left the water.

Pursuing its course, the stream, sometimes gradually, sometimes suddenly by means of a waterfall, finds its way from the hills to the flatter ground below, and changes from a turbulent mountain torrent to a slower running river. The reduction in the force of the current enables suspended particles to be deposited on the bottom, particularly at bends in the river, and plants, such as filamentous algae, are now enabled to grow attached to the stones or gravel. As the river nears its mouth more and more suspended matter is deposited, permitting a wealth of higher plants to take root, particularly at the margins, and they themselves cause further accumulations of silt. Thus are formed stiller, weedy shallows along the banks with a flora and fauna very similar to those of a pond or ditch. Here the nymphs of dragonflies crawl about the bottom, and climb up the plants out of the water when about to emerge as adults. Here, too, many

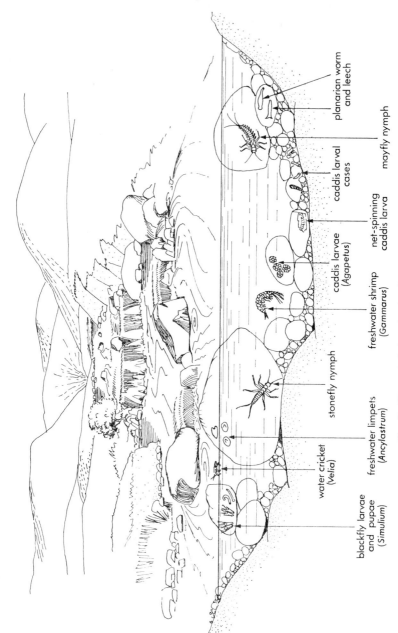

planarian worm
and leech

mayfly nymph

caddis larval
cases

net-spinning
caddis larva

caddis larvae
(Agapetus)

freshwater shrimp
(Gammarus)

stonefly nymph

freshwater limpets
(Ancylastrum)

water cricket
(Velia)

blackfly larvae
and pupae
(Simulium)

Fig 8   Characteristic animals of a stony stream

caddis larvae, beetle larvae, and other creatures associated with still water occur, while in sandy lagoons caused by the meandering of the river, the *burrowing* nymphs of mayflies will be found, often in huge numbers. In the mud, aquatic worms will be numerous, and perhaps freshwater mussels. Freshwater sponges may form encrusting growths on submerged objects. (Plate **2**)

Where the river reaches the influence of tidal action the life becomes sparse, for few forms can survive the daily changes of salinity. A few, however, can do so, and among these are one or two species which are of especial interest, for they seem to be in the process of penetrating pure fresh water.

Although not all rivers will be like the one whose course has been traced, in general there will be a steady gradation, similar to that which has been followed, from one type of habitat to the next. The conditions of life in running water have been very inadequately studied, and here again, as in pond life, there is a promising field awaiting those who are prepared for systematic work.

This brief review of the three main types of aquatic environment—lakes, ponds and rivers—has not taken into account the infinite variations that are met with in them all, and it has been possible to consider only examples of each. Other aquatic habitats such as fens, bogs, marshes, springs, subterranean waters and many more all have their individual characteristics, and will provide endless scope for detailed study of their inhabitants. Canals, especially in their disused stretches, are often particularly rich in both plant and animal life. Apparently unpromising habitats such as the rain-water tubs in the garden, bird-baths, cattle drinking-troughs, pools of water in holes in trees, or even the rain-water gutters of houses, should not be ignored by the student. Each of these is a profitable source of living material and often the only one for a particular group of organisms.

In all these habitats will be found characteristic communities of plants and animals, with intricate ecological relationships which will repay closer study.

The suggestion that an inland body of water could be regarded as a closed community seems to have been made by the American biologist S. A. Forbes who, in a classic publication on limnology, *The Lake as a Microcosm*, a paper published in 1887, wrote of a lake:

> It forms a little world within itself—a microcosm within which all the elemental forces are at work and the play of life goes on in full but on so small a scale as to bring it easily within the mental grasp.

No bodies of inland water are strictly closed communities, of course, but they are probably the nearest approach to this ideal that can be found in nature.

Fig. 9 shows the main features of the energy flow in such an ecosystem. The first constituents are the energy of sunlight, gases from the atmosphere, and inorganic matter dissolved in the water. These can be converted by green plants and some bacteria into living substances on which all other organisms, plant and animal depend. On their death, all living things disintegrate and are converted back into detritus and dissolved substances, and are thus returned to the common pool of nutrients. Some organisms, such as the adult stages of insects and amphibians, are lost temporarily to the ecosystem.

Fig. 9   Energy flow in fresh water

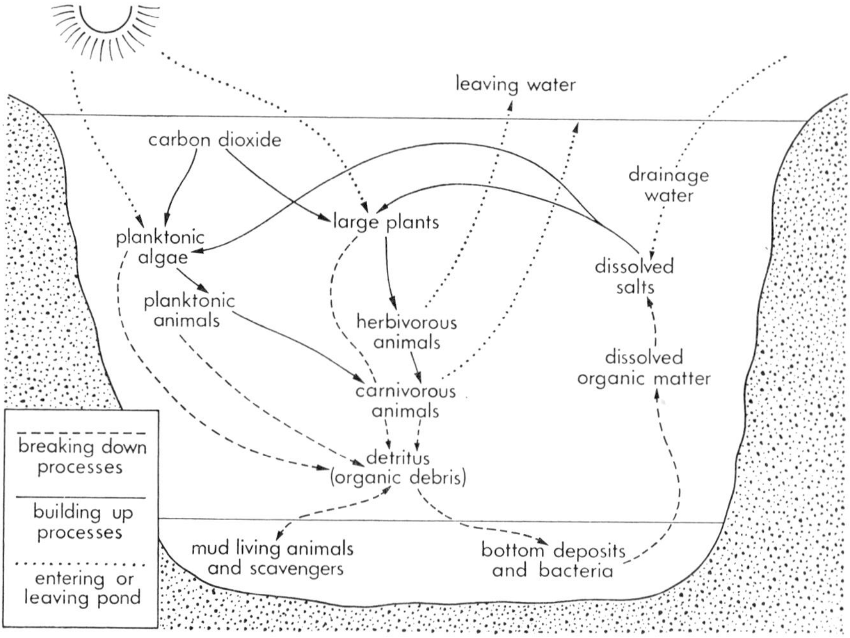

There are thus two main groups of processes at work: a building up of simple food materials to more complex substances, and a breaking down by bacteria of these to simpler nutrients. It need hardly be said that this very brief account is a considerable over-simplification of what is, in reality, a very complex series of processes.

3: PLANTS OF THE WATERSIDE AND THE LARGER AQUATIC PLANTS

The aquatic and waterside vegetations have, from the earliest times, been fruitful sources of inspiration to poets and artists, and few of us, however unromantic, can remain unmoved by the continuous pageant of wild flowers which add colour and beauty to our ponds, ditches and streams from early spring to late autumn.

The very luxuriance of these plants indicates that the water and the waterside offer special advantages as a place in which to live, and many representatives from widely differing families of flowering plants are found growing in one or other of these habitats.

What, then, are the advantages of life in or near the water? First and foremost is the lessening of the danger of drought, a calamity which must often befall terrestrial plants not in close proximity to aquatic habitats. Secondly, nutrients are usually plentiful in an easily available form; while for plants entirely submerged, temperature changes are not so violent as on land.

Nevertheless, there are certain disadvantages. For instance, pollination of the flowers is difficult for totally submerged plants, and perhaps this is why many of them propagate by vegetative means. In waterlogged soils, or in muds, the supply of oxygen to the roots may be very scarce. As we shall see, however, such difficulties are usually offset by structural modifications of the plants concerned.

It is these adaptations to the special conditions of an aquatic existence that makes the study of water-plants of such absorbing interest. In those that are almost, if not entirely, submerged, the water acts as an all-round support, dispensing with the need for strengthening tissue to keep them erect, and so, on the whole, they have fragile, delicate stems and branches. The epidermis, having no cuticle, allows absorption of water, dissolved gases and other nutrients to take place all over the submerged surface of the plants.

Stomata, the small pores on the leaves of plants through which excess water vapour passes out, are absent from the leaves of water-plants, except in those species such as water-lilies whose leaves float on the surface. In these cases, stomata occur on the upper side only.

Water-plants need the same substances for growth as do their relatives on land, but by the nature of their environment they get them in rather a different way. Food materials such as nitrates, carbonates, phosphates and sulphates are usually plentiful in a dissolved state in the water, and these are absorbed, not only by the roots, as is usual, but also all over the submerged surface of the plant through the thin epidermis.

Carbon dioxide, needed for photosynthesis, as well as oxygen, which is just as essential to plants as to animals, are both obtained in solution from the water.

Oxygen is not plentiful in water even in solution, and in the mud it may be absent altogether. To enable respiration to be carried on, aquatic plants tend to store air, often in large spaces within the stems, leaves or roots, and these air spaces frequently link up the various parts of the plant to ensure adequate oxygenation for all the tissues.

Most water-plants are perennials, and these vary in their methods of tiding over the winter. Some, like the starworts, *Callitriche* (Plate 4, 2) remain relatively unchanged and merely sink to the bottom of the water until spring. Others, such as the water-lilies, have stout perenniating stems called rhizomes (Fig. 11) buried in the mud, in which food in the form of starch is stored, and from which the new shoots arise in the spring. Still others, such as frog-bit, produce in the autumn, special shoots called turions or 'winter-buds', similarly packed with reserve food; these sink to the mud on the decay of the old plant and remain there until spring.

Even a cursory glance at the vegetation in and around a pond will reveal the presence of several distinct communities of plants, each more or less confined to a particular zone. Although there may be a little wandering of some types into a neighbouring zone, the plants of a typical 'lowland' pond may be separated in the following way:

Zone 1   Plants living in the marshy area some distance from the water.

Zone 2   Those living in swampy conditions immediately at the edge of the water.

Zone 3   Plants rooted in the mud in the shallow water at the margins, but having leaves either floating on the surface or standing out of the water.

Zone 4   Plants floating freely at the surface with no roots attaching them to the bottom.

Zone 5   Those farther out in deeper water, rooted in the mud with their vegetation completely submerged.

Conditions may vary a great deal between pond and pond, and one or other of the zones may be absent. Then again the 'hardness' of the water may determine which particular plants are present or absent. Thus the water soldier, *Stratiotes aloides*, favours alkaline water, while quillwort, *Isoetes lacustris*, can thrive in acid waters poor in nutrients.

It is obviously impossible in one chapter to discuss all the plants found in each zone, and only a few typical representatives can be mentioned.

Marsh Plants   The plants in the marshy area surrounding the pond are, of course, not truly aquatic, but they exist in a habitat which is nearly always moist.

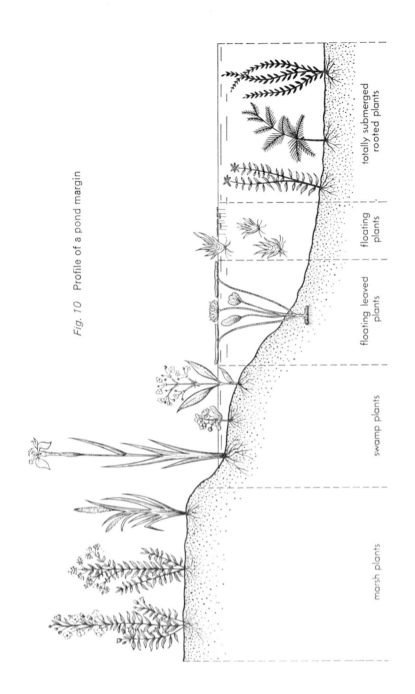

Fig. 10 Profile of a pond margin

marsh plants

swamp plants

floating leaved plants

floating plants

totally submerged rooted plants

They are usually referred to as *hygrophytes* (Gr. *hygros*, damp; *phtyon*, plant), in contrast to the true water-plants, which are called *hydrophytes* (Gr. *hydro-*, water). The waterlogged soil in which they live has water occupying the space between the soil particles instead of air, and it is therefore deficient in oxygen. On the other hand, it is usually rich in mineral salts. The marsh vegetation is lush, and the typical members of it often have large leaves (since they have no need to economize moisture), and a spongy structure plentifully supplied with aerating tissue for the storage and passage of air.

The lovely marsh marigold or kingcup, *Caltha palustris*, is a typical member of this community, with its large kidney-shaped leaves and sturdy root-stock, which holds the plant firmly in the soft ground. The large, showy yellow flowers, which are great attractions for insects, are among the first to deck the pond-side in spring. It is, perhaps, not generally realized that the marsh marigold flowers have no petals; it is the sepals alone which make up the flower, changing in colour from green through yellow to a beautiful rich gold as they develop. (Plate 8)

Other plants of the marsh are great willow-herb or codlins-and-cream, *Epilobium hirsutum* (Plate 8), whose rose-coloured flowers and tall growth make it easy to recognize; meadow-sweet, *Filipendula ulmaria*, whose many country names—pride of the meadow, meadow queen, honey-flower, queen's feathers and a host of others—show the regard in which its feathery clusters of sweetly fragrant flowers have been held by country people (Plate 8); purple loosestrife, *Lythrum salicaria*, whose tapering spires of purple flowers are such a feature of ponds and ditch-sides in late summer in some parts of the country (Plate 8); the bur-marigolds, *Bidens cernua* and *B. tripartita*, with somewhat inconspicuous flowers, but with two-spined achenes or 'fruits' which attach themselves most securely to trousers or stockings (Plate 8); yellow iris or flag, *Iris pseudacorus*, whose most handsome flowers and sword-shaped leaves add charm to the scene; the exquisite water forget-me-not, *Myosotis scorpoides*; and finally the various species of rushes, *Juncus*, and sedges, *Carex*.

Although scarcely marsh plants, alder trees and various species of willow are often found at the margin of ponds, their leaves in summer providing shade over the water, their trunks, convenient resting and transformation places for the aerial stages of many aquatic insects, and their submerged rootlets anchoring places and refuges for a host of aquatic creatures. Landward of the marsh plants around well-established ponds and some lakes, the ground is colonized by bushes of various kinds. The name 'carr' is given to this vegetation zone. Farther away from the water still, trees, including oak, may occur.

Swamp Plants The marshy area merges almost imperceptibly into the swampy region of the pond, where the ground is covered with water at all except the driest times of the year. The plants here are generally tall, and often have long creeping rhizomes which, by taking a firm grip of the mud, keep the plant upright. It is characteristic, too, of some of them that they produce abundant buds on the rhizomes, each of which gives rise to a new plant, so that in quite a short time the rhizomes form a solid mat, and a dense community of one species comes into being, effectively crowding out all other plants. The reed, *Phragmites communis*, is a notable member of this group, and in some parts of the country forms dense reed-beds round the edges of ponds or lakes. The so-called bulrush or great reedmace, *Typha latifolia*, and the true bulrush, *Schoeno-plectus lacustris*, are also familiar plants in such situations, and similarly form dense beds extending into the open water.

The decay of vegetation and the collection of silt round the base of such plants will, in time, raise the level of the substratum, and may, in fact, result in the drying-up of the water, leaving only a marshy area to show where once there had been a pond.

Other familiar plants here are the bur-reeds, *Sparganium*, best known, per-haps, when their round spiky balls of fruit are in evidence in late summer, water plantain, *Alisma plantago-aquatica*, and the beautiful flowering rush, *Butomus umbellatus*. At the edge of the water will be seen the densely packed rose-red flower spikes of amphibious persicaria, *Polygonum amphibium*, with the elongated leaves floating on the surface.

Fig. 11   Rhizome of yellow water-lily, *Nuphar lutea*

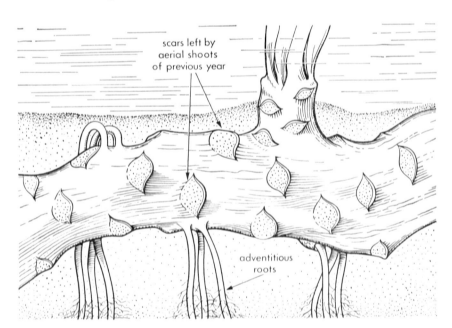

scars left by
aerial shoots
of previous year

adventitious
roots

True Aquatics Rooted in the Mud in the Shallows   So far plants that could be described as semi-aquatic have been considered, as they do not need to grow in standing water, but thrive quite well provided their environment is damp. The remaining plants mentioned in this chapter are the true aquatics, and thrive only when growing actually in the water.

In the first group of true aquatic plants, those growing in the shallows and rooted in the mud, yet with their leaves either floating or standing above the surface, are included the most beautiful of all water-plants the water-lilies; two species, the white, *Nymphaea alba*, and the yellow, *Nuphar lutea*, are common in this country. They have stout rhizomes buried in the mud (Fig. 11), and from these arise branches bearing the large floating leaves. The leaf stalks are generally at an angle, but become more upright as the water-level rises, thus keeping the leaves resting on the water. The upper surface of the leaves is coated with a waxen layer off which the water runs easily, and it bears stomata. The underside of the leaves has no stomata and, being thin-walled, permits absorption of gases and salts from the water. Large air spaces occur in the leaves and are continuous with air passages running down the leaf stalks to the submerged stems with oxygen.

The flowers are borne above water and are pollinated by insects. The bottle-shaped fruits of the yellow water-lily (which have earned for the plant its country name of brandy-bottle) break away from the plant and float about on the surface of the water until they disintegrate, when the individual seeds sink to the bottom. The fruit of the white water-lily sinks to the bottom and eventually bursts, releasing as many as two thousand seeds which float for a few days on the surface, and spread widely before sinking again. In autumn the plant dies down, its accumulated food reserves being stored in the rhizome, from which in spring arise new shoots. Superficially resembling a small yellow water-lily is the so-called fringed water-lily, *Nymphoides peltata*, although it belongs to a different family, Menyanthaceae. It has five yellow, fringed, petal-like corolla-lobes and small, rounded, toothed leaves which are purple below. It is somewhat local in distribution, being found in quiet waters in southern England, but it appears to be spreading. The other and more familiar member of the family is bogbean, *Menyanthes trifoliata*, whose spikes of white and pink flowers are conspicuous in early summer in swamps and shallow waters.

The water crowfoots, *Ranunculus* spp.,[1] which in spring cover the whole surface of many ponds and streams with their white flowers, are interesting as examples of water-plants possessing two distinct kinds of leaves—floating leaves which are flat and rounded, and submerged leaves which are finely divided into a large number of hair-like filaments. Various suggestions have been put forward to account for such differences in leaf structure. Some botanists have considered that the finely dissected leaves have the advantage of presenting a

---

[1] The abbreviation *spp.* indicates that several species are being considered.

Pl. 1

Above: Ennerdale Water, Cumberland: an oligotrophic lake. *Below:* Esthwaite Water, in the Lancashire part of the Lake District: a eutrophic lake. (See page 22)

Pl. 2
    Above: The River Duddon, Cumberland: a fast river. *Below:* The River Stour, near Flatford Mill, Suffolk: a slow river

Pl. 3

Above: The old brick-pond, Wicken Fen, Cambridgeshire. *Below:* Cunsey Beck, near Windermere

Pl. 4

AQUATIC PLANTS

1 Canadian pondweed, *Elodea canadensis*. 2 Water starwort, *Callitriche stagnalis*. 3 Water milfoil, *Myriophyllum spicatum*. 4 Hornwort, *Ceratophyllum demersum*. 5 Water violet, *Hottonia palustris*. 6 Water crowfoot, *Ranunculus aquatilis*

Pl. 5

Frog-bit, *Hydrocharis morsus-ranae. Above:* Floating on the surface of the water; two 'winter-buds' are seen developing on lateral stolons. *Below:* Three 'winter-buds' at different stages of development floating at the surface in spring

Pl. 6

Water soldier, *Stratiotes aloides*. *Above:* Floating at the surface in summer; the white flower can be seen in the specimen in the middle foreground. *Below:* The same plants in late summer

Pl. 7

1 Water moss, *Fontinalis antipyretica* (× ⅓). 2 *Sphagnum*, or bog moss (× 1). 3 Lesser duckweed, *Lemna minor*; on the right are a few thalli of greater duckweed, *L. polyrrhiza* (× ½). 4 Water-fern, *Azolla filiculoides* (× ½)

Pl. 8

MARSH PLANTS

1 Great willow-herb, *Epilobium hirsutum*. 2 Meadow-sweet, *Filipendula ulmaria*. 3. Marsh marigold, *Caltha palustris*. 4 Bur-marigold, *Bidens cernua*. 5 Great marsh sedge, *Carex riparia*. 6 Soft rush, *Juncus effusus*. 7 Purple loosestrife, *Lythrium salicaria*

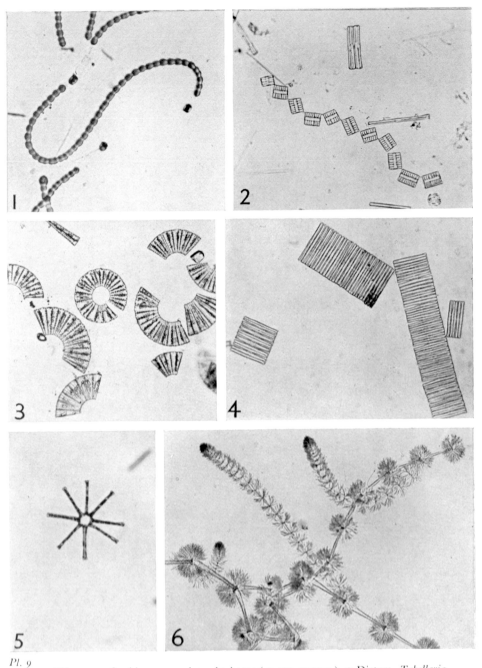

Pl. 9   1 Filaments of a blue-green alga, *Anabaena* (× 500 approx.). 2 Diatom, *Tabellaria flocculosa* (× 120). 3 Diatom, *Meridion circulare* (× 250). 4 Diatom, *Fragilaria capucina* (× 175). 5 Diatom, *Asterionella formosa* (× 150). 6 Frog-spawn alga, *Batrachospermum moniliforme* (× 20)

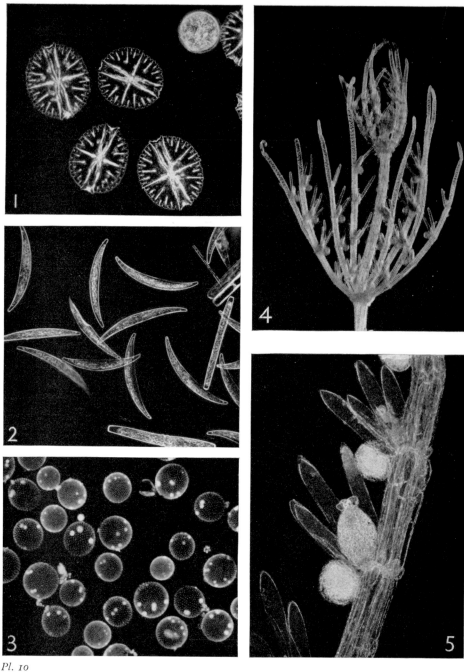

Pl. *10*

1 Desmid, *Micrasterias* (× 50 approx.). 2 Desmid, *Closterium* (× 125 approx.). 3
Volvox (× 20). 4 Stonewort, *Chara fragilis:* end of main shoot showing reproductive
bodies (× 7). 5 *Chara fragilis:* the pear-shaped female oogonium and the round male
antheridium (× 50)

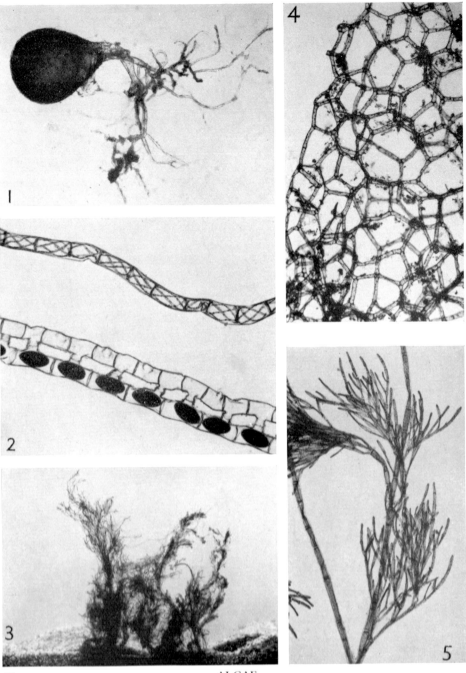

Pl. 11 ALGAE

1 *Botrydium granulatum* (× 45). 2 Filaments of *Spirogyra* (× 75). *Above:* Normal filament. *Below:* Two filaments in lateral conjugation; the black objects are zygospores. 3 *Cladophora* (× ½). 4 Water-net, *Hydrodictyon reticulatum* (× 5). 5 Filaments of *Cladophora* (× 15)

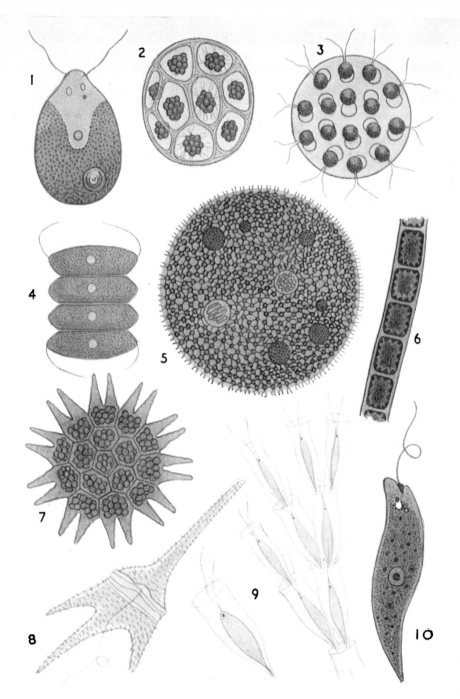

Pl. 12   PLANKTONIC ORGANISMS

1 *Chlamydomonas* (× 1,500). 2 *Pandorina* (× 600). 3 *Eudorina* (× 600). 4 *Scenedesmus* (× 1,500). 5 *Volvox* (× 150) showing two antheridia and a number of oogonia. 6 *Microspora* (× 500). 7 *Pediastrum* (× 950). 8 *Ceratium* (× 375). 9 *Dinobryon* (× 825). 10 *Euglena* (× 450)

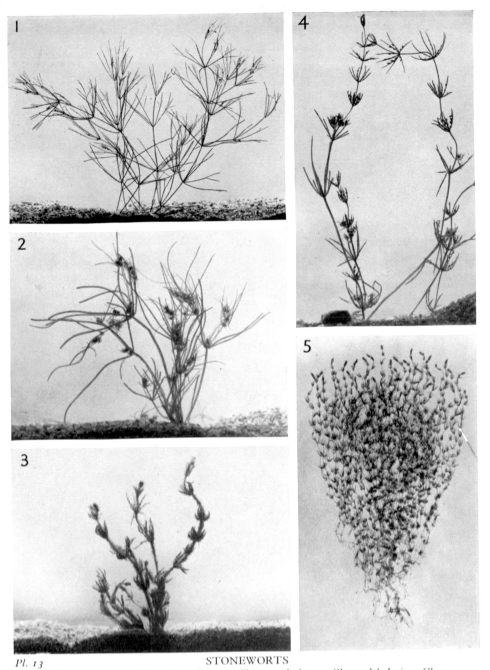

Pl. 13   STONEWORTS

1 *Nitella flexilis.* 2 *Tolypella prolifera.* 3 *Chara aculeolata.* 4 *Chara globularis.* 5 *Chara aspera* (mounted as a herbarium specimen on paper)

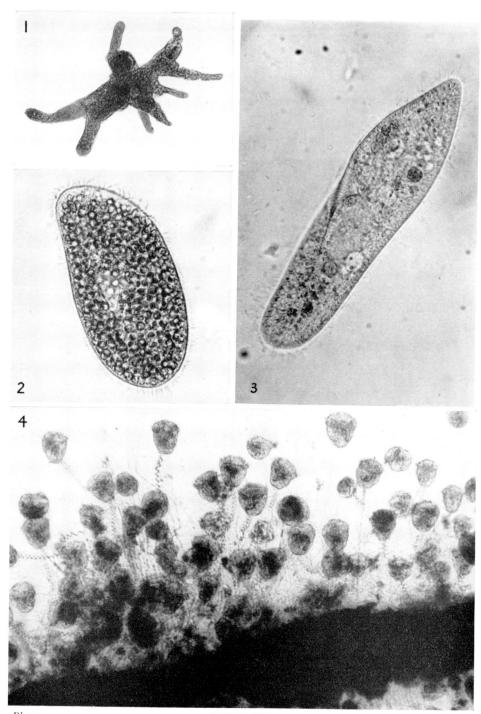

Pl. 14

PROTOZOA

1 *Amoeba* (× 150). 2 *Paramecium bursaria* (× 400). 3 *Paramecium caudatum* (× 300).
4 *Vorticella*, colony (× 100)

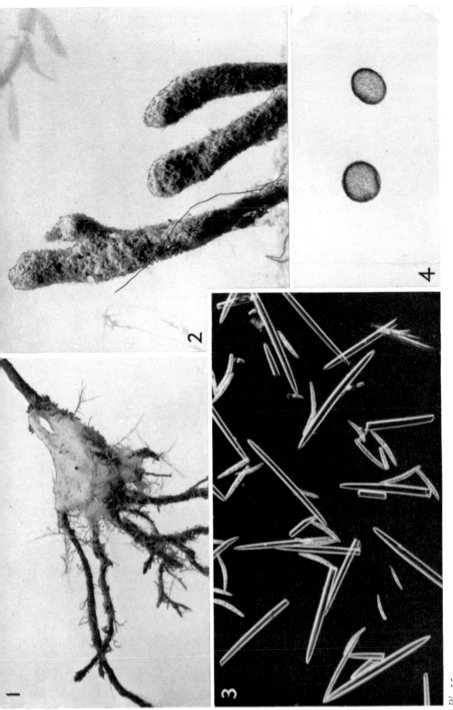

1 A small growth of freshwater sponge, *Ephydatia fluviatilis*, on alder tree rootlets (× ½). 2 Finger-like growths of *Euspongilla lacustris* (× ½). 3 Spicules of *Euspongilla lacustris* showing the smaller, rough-surfaced spicules among the longer ones (× 120). 4 Gemmules of freshwater sponge (× 25)

Pl. 15

Pl. 16

1 Colony of *Cordylophora lacustris* (× 15). 2 Hydra with testes (at the top of the column) and ovary (× 20). 3 Hydra: tentacle tips showing the epidermal cells containing nematocysts (× 70 approx.). 4 Hydra with bud

larger assimilating surface to the water than if they were entire, and others have suggested that this type of leaf may be of value by offering less resistance to water currents. Neither explanation seems adequate, and recent research indicates that the leaf-form is a response to differences in the type of food available to the plant during the growing season, and in particular to the relative proportions of carbohydrates and mineral salts. At the beginning of the season the dissected form of leaf is produced, but as the plant grows to the surface and the carbohydrate content of the plant increases, the floating leaves appear. In support of this theory it has been found that when these plants are kept artificially under a high light intensity, but at a low temperature (conditions favourable to carbohydrate accumulation), the floating type of leaf can be produced under water. (Plate 4, 6)

The arrowhead, *Sagittaria sagittifolia*, a common plant in many waterways in the country, goes one stage farther than the crowfoots, for it can bear *three* kinds of leaf—a submerged, grass-like type, which is the first to appear in spring; an ovate, floating type, which follows later; and finally, the characteristic, arrow-shaped leaf, which stands up above the water and from which the plant derives both its English and Latin names. Specimens of arrowhead, grown under shady, warm conditions (favourable to the accumulation of nitrogenous substances rather than carbohydrates), continue to produce the grass-like leaves indefinitely.

The pondweeds, *Potamogeton* spp., are a large family, and it is by no means easy to distinguish between some of the species. A number are completely submerged plants, but the broad-leaved pondweed, *P. natans*, perhaps the commonest, has floating leaves. The rhizomes of this plant form dense mats at the bottom of the water, and a forest of shoots rises from them, some spreading more or less horizontally in the water, and others, more erect, bearing the large floating oval leaves. In common with the leaves of water-lilies, these serve as platforms for many aquatic creatures. The egg masses of snails and of some aquatic insects, colonies of bryozoa, and the curious oval cases of the larvae of the china mark moths are some of the interesting objects which are revealed by an examination of the underside of these floating leaves.

Submerged Plants The plants which are totally submerged in the water and rooted in the mud form, perhaps, the largest group. Of necessity, the whole of their oxygen and carbon dioxide is absorbed from the water. A good proportion of their other nutrients is similarly taken from the water by absorption, but the roots of aquatic plants, where present, do absorb mineral salts from the mud just as those of land plants do from the soil. The roots are not merely anchors as was formerly believed.

The species in this group are so numerous and their characteristics so diverse, that it is difficult to know which to mention as typical representatives.

Perhaps the most abundant of all our aquatic plants is the Canadian pondweed, sometimes more picturesquely called water thyme, *Elodea canadensis*. (Plate **4**) It is such a feature of almost any stretch of stagnant or slow-running water, that it is hard to believe that it was unknown in the British Isles before 1836, when it was recorded in County Down, in Northern Ireland. It is characteristic of this plant that on introduction into a new and favourable habitat, it multiplies at an incredible rate by simple vegetative means, and rapidly chokes the water. Then, in a few years, its vigour seems to wane. It has been suggested that this phenomenon is due to the depletion of some nutrient or other growth factor in the water. A few plants of this North American species introduced into a tributary of the River Cam in 1848 had not only spread to the main river within four years, but had so completely choked it as to make boating, fishing and swimming impossible. About the same time, it gained access to the drainage dykes in the Fens, and so impeded drainage that a Government adviser was called in to help tackle the problem. In spite of all the steps that were taken, *Elodea* flourished unchecked for a number of years, and then, without any apparent cause, its luxuriance waned. Although still an abundant plant, its peak can be said to have occurred in this country between 1850 and 1880, and it has now taken its place as a normal inhabitant of our stiller waters.

Although male specimens of *Elodea* have been recorded in Britain, only the female plants are commonly found. The amazing luxuriance is, therefore, entirely due to vegetative reproduction—by the breaking of the extremely brittle stems, each fragment of which grows into another plant. Thus, every *Elodea* plant in the country can be considered as a surviving part of the original plants that were introduced about 1840. *Elodea* survives the winter by developing 'winter shoots', consisting of tightly wrapped leaves, in the autumn.

The water milfoils, *Myriophyllum* spp., are attractive plants with delicate feathery leaf-whorls along the length of the stems. The leaves are submerged, but in the spiked milfoil, *M. spicatum*, the minute flowers, arranged in small clusters, are borne in a slender spike about 50 mm long which grows above the surface, from the top of the plant. In the whorled milfoil, *M. verticillatum*, however, the flowers are submerged and borne in the axils of the upper leaves, although if the plant is growing in very shallow water this plant, too, may have its flowers above the surface. (Plate **4, 3**)

In autumn, and at other times when conditions are unfavourable, the milfoils produce sausage-shaped dark-green turions, or winter-buds, which remain dormant until spring, when they expand into graceful new plants.

Floating Plants The last group of aquatic plants to be considered are those that float freely in the water and are not rooted in the mud at all. They usually do have roots, but these hang loosely, serving not only for the assimilation of water and its dissolved substances, but also as balancing organs.

Everyone will have seen how the duckweeds, or 'pond scum', *Lemna* spp., often cover a pond with a continuous carpet of green. They are only tiny plants, yet they grow in such a dense aggregation that their minute 'leaves', or more correctly thalli, can completely hide the water below. I well remember in my childhood walking by mistake into a duckweed-covered pond, thinking it was dry land, and the memory of a dripping return home has effectively prevented me from making the same mistake again!

From the centre of the underside of the thalli the roots hang down into the water. Budding takes place throughout the spring and summer, new thalli arising from slits in the sides of the old ones, from which they eventually break away. The thalli have stomata on their upper surface and can thus absorb oxygen and carbon dioxide from the air, while their under surface and the roots absorb these gases and salts dissolved in the water.

Duckweeds are often found in very foul water, such as ponds fed from farm-yard manure heaps, in which other flowering plants cannot survive. In fact, it is believed that a high organic content in the water is necessary for some species of *Lemna* to thrive. In my experience, ponds covered with duckweed are rarely

Fig. 12 Ivy-leaved duckweed, *Lemna trisulca*.
Right: thalli enlarged

productive of much animal life, and this is probably because the type of pond that best suits duckweed is too deficient in oxygen for animal life to thrive in it.

The commonest species in this country is the lesser duckweed, *Lemna minor* (Plate 7, 3), whose thalli measure from 1·5 to 4 mm across. *Wolffia arrhiza* has the distinction of being our smallest flowering plant, its thallus being about the size of a small pin's head. In contrast, the greater duckweed, *L. polyrrhiza* (Plate 7, 3), has thalli about 5 mm to 8 mm across, rather like miniature water-lily leaves.

The ivy-leaved duckweed, *L. trisulca*, differs from the rest of the species in living totally submerged, just below the surface, obtaining all its gases and salts from the water. The individual thalli, about 8 mm across, are not really ivy-shaped but are elliptical, and each bears an elongated portion resembling a stalk.

The ivy-leaf appearance occurs when two new thalli are developing from one old one. A minute green alga, *Chlorochytrium lemnae*, lives in the intercellular spaces of this species, but seems to derive no benefit from the association other than a place in which to live.

The tiny flowers of the duckweeds are borne above the water-level and are believed to be pollinated by insects. By the autumn, duckweed thalli have accumulated starch reserves and they sink to the bottom under their weight, remaining dormant through the winter, although in sheltered places a few plants can usually be seen floating even at that time. In spring, as the starch becomes used up, the thalli become lighter and float to the surface again. Some species, such as *L. polyrrhiza*, produce special thalli in autumn; these are smaller than the normal ones, dark in colour, well-packed with starch and have very few air spaces. In *L. gibba* the lower part of the thallus is much swollen at all times of the year and is a mass of spongy tissue. In winter the thalli of this species may take root in the substratum.

Frog-bit, *Hydrocharis morsus-ranae*, is another plant that sometimes covers the entire surface of a small pond or ditch. Its leaves are very similar to small water-lily leaves, and grow in the form of a rosette. The roots hang down in the water and, when many plants are growing together, are so closely matted that if one plant is removed from the water, a large bunch comes with it. The three-petalled white flowers are borne freely, but seed is very rarely set in this country and propagation is vegetative. The winter buds of frog-bit are perhaps the best known of all winter buds, and are used in schools and colleges to demonstrate this feature of plant life. Throughout spring and summer, leaf buds are formed on lateral shoots, or stolons, but as autumn approaches, the winter buds, which are different in structure from the usual buds, begin to form on stolons hanging down in the water.

These contain, in addition to the embryo plant, food reserves in the form of starch and, when ripe, they break off from the stolon and sink to the bottom of the water. Here they remain dormant until spring, when the using up of the starch reserves and the development of air spaces makes the winter buds lighter and they float to the surface to develop into small plants. (Plate 5)

It would hardly be supposed that frog-bit was related to Canadian pondweed, *Elodea*, as they seem to bear no resemblance to each other. Yet, they are included in the same family, Hydrocharitaceae, while another member of this family, the water soldier, *Stratiotes aloides*, is again quite different in superficial appearance from either. This curious plant is local in its distribution, although common in the fens of eastern England. I can remember my surprise, some years ago, on being shown a pond literally full of this plant on a housing estate in an industrial area of Lancashire. Presumably it had been introduced by some local botanist, and it was obvious from the manner in which it had multiplied that it had found a suitable environment.

The name water soldier refers to the very sharp sword-like leaves which arise from a central short, but stout, stem. The specific epithet *aloides*, however, gives a better clue to its form, for in appearance the plant closely resembles an aloe.

The plant is normally totally submerged just below the surface, but it monopolizes its stretch of water and no plants grow immediately above it. From the lower parts of the stem a number of straight roots hang down into the water, and may terminate in a pad-like growth which reaches to the bottom mud. These roots are of great importance in maintaining the equilibrium of the plant. Not only does it lose its erect position if these are cut off, but it is almost impossible to maintain the plant on an even keel if it is transferred entire to another stretch of water of a slightly different depth.

In spring the plants rise to the surface so that their leaves are well out of the water. The white flowers then appear, but in this country seed is not set, as the plants are all females. When flowering is over, the plants sink and become once again totally submerged. (Plate 6) The water soldier favours waters rich in calcium bicarbonate and the abstraction of carbon dioxide from the water by the plants during photosynthesis sometimes causes a white deposit (chalk or calcium carbonate) on the leaves.

One of the most interesting groups of aquatic plants, and members of this floating flora, are the bladderworts, *Utricularia*. They are commonest in waters rather poor in mineral salts, such as peaty pools, and are found floating totally submerged just below the surface. There are no true roots, but the branches themselves are root-like, consisting of innumerable hair-like leaves on which small bladders are borne. Each bladder has a nearly circular aperture at one end, around which slender bristles form a kind of funnel closed by a trap-door opening inwards. The inner surfaces of the bladders bear hairs which absorb the liquid inside, and when the trap-door is shut, the abstraction of the internal liquid causes a contraction of the bladder under tension. When small creatures, attracted it is believed by mucilage glands near the entrance to the bladder, touch the sensitive bristles around the aperture, the trap-door is released and water rushes into the bladder, carrying the creatures with it. The 'trap' then closes, and in due course the animals die and their decomposed remains are absorbed by the plant. In this way deficiencies in food materials in the waters frequented by the bladderworts are made up.

Bladderworts reproduce by vegetative means, and also by producing fruits. The yellow flowers are borne on shoots which grow out of the water. In late summer and autumn 'winter buds' are produced and, when the parent plant dies down and sinks, the buds are taken down and remain there until spring, when they become detached from the now disintegrated plant, and float to the surface to develop into new plants.

The last plants to consider in this group of floating aquatics are perhaps the ones most completely adapted for living in water. These are the hornworts, *Ceratophyllum demersum* and *C. submersum* (Plate 4), which not only live entirely submerged at all seasons, but flower under water, pollen being conveyed to the stigmas through the water. The male and female flowers are separate, but are borne on the same plant. They are inconspicuous, and grow in the axils of the leaves. The stamens of the male flowers become detached, floating to the surface of the water, where they dehisce. The pollen sinks, and thus comes into contact with the stigmas of the female flower.

In the hornworts there is an entire absence of true roots, although at times lightish-coloured shoots are developed which penetrate the mud. In general, however, the plants float freely in the erect position in the water. In spring the new shoots are near the surface, but they tend to sink near the bottom as they get older, and often form dense jungles of vegetation in favourable conditions, providing shelter and support for innumerable small creatures such as insects, worms and snails.

Here, then, are plants which have become completely adapted to an aquatic life, severing connection with the soil, and not only deriving gases and salts from the water, but also effecting pollination in the same medium.

Flowerless Plants

The plants which have been discussed so far in this chapter have been flowering plants. Before passing on to the next chapter to deal with the smaller non-flowering plants such as the algae, it will be convenient to mention here one or two of the non-flowering plants which are sufficiently large to merit inclusion in the scope of the present chapter. These are the mosses and ferns.

Of the mosses, the most important are the species of bog mosses, *Sphagnum* (Plate 7, 2), which form soft carpets of vegetation in the damp areas surrounding some ponds and streams, and even invade the water itself, supported on the entangled roots of other plants. The dead remains of these plants have helped to form the vast deposits of peat which exist in many places. The stems and leaves of bog mosses have, between the normal cells making up their structure, cells which are devoid of protoplasm and serve merely for the absorption and storage of water. Thus it is that these plants can absorb twenty times their own weight of water, a property which has enabled them to be used for emergency wound dressings in times of war.

A number of mosses grow on the rocks and stones in streams, and provide shelter for an interesting population of insects, rotifers, Protozoa and other creatures that would otherwise be unable to live in such rapidly moving waters. Far more imposing than these, however, are the true water mosses of the genus *Fontinalis*. *F. antipyretica* (Plate 7, 1) is the only common species and occurs in both running and still waters, with stems sometimes up to one metre in length, which are usually attached to stones or tree stumps. The spore-bearing capsules are rarely produced when the plant is totally submerged, but they are often produced in great numbers if the water dries up sufficiently to leave the stems high and dry.

While there are no native water-ferns, *Azolla filiculoides* (Plate 7, 4), sometimes referred to as 'fairy moss', although it is a true fern, has been introduced to this country from the United States as a plant for aquaria and garden pools, and has spread to natural waters in some parts of the country. No doubt it will establish itself in time as *Elodea* has done. It is a floating plant, in appearance like a thick-fronded fern, but only about 15 mm across. Each frond consists of two lobes, one of which is floating and the other submerged; the latter assists in the absorption of water and dissolved nutrients. Within a small cavity of the upper lobe, which has a narrow orifice opening upwards, a blue-green alga, *Anabaena azollae*, is usually found, which is able to fix atmospheric nitrogen.

The colour of the fronds varies greatly from a pale green to bright red. Since *Azolla* spreads very rapidly on waters that suit it, and soon covers the surface with a thick mat of vegetation, it tends to become a pest, but as it is not hardy, a severe winter can kill it off.

In this chapter the plants have intentionally been discussed in the order in which they would be discovered by the reader when visiting a freshwater habitat, starting with those on relatively dry land and proceeding towards, and into, the water. A botanist, however, considers the succession of plants—called the hydrosere (L. *sero*, to plant, put in a row)—in the reverse direction, starting from deep, open water and finishing on land, thus: completely submerged plants, floating species, those rooted in the mud but with floating leaves, swamp plants (the so-called 'reedswamp community'), marsh plants and, finally, carr and woodland.

The sequence of plants in a hydrosere reflects the degrees of adaptation to an alien environment of formerly terrestrial plants in their adoption of aquatic conditions. The classic hydrosere is seen at its best in smaller, sheltered bodies of water, particularly where an inflowing stream deposits silt. In a pond the plant zones tend to form concentric circles; in ditches, canals and slow stretches of rivers they are in linear zones parallel to the bank. Not all zones are invariably present and there may be some inter-mixing of plants into neighbouring zones.

The value of plants in an aquatic habitat is as important as on land. Their ability to build up, by photosynthesis, organic material from inorganic substances provides crops of potential food for direct or indirect exploitation by aquatic animals. The oxygen produced, and the carbon dioxide absorbed in the process, are valuable to the economy of the habitat.

The stems and leaves of water-plants provide mechanical support, as well as shelter from predators and excess sunlight. They serve as places on which to deposit eggs, and some animals pass their early stages within the plant tissues.

Finally, the death and decay of plants, especially those of the reedswamp community, can completely change the character of a body of water, even to the extent of causing it to disappear altogether—the usual fate of a smaller stretch without man's active or unwitting intervention.

An intriguing aspect of aquatic plants is that many species are very widely distributed geographically, some throughout the world. The latter include hornwort, *Ceratophyllum demersum*, and lesser duckweed, *Lemna minor*. Others that are widespread are reed, *Phragmites communis*, the pondweeds, *Potamogeton crispus* and *P. pectinatus*, reedmaces, *Typha angustifolia* and *T. latifolia*. There has been some speculation as to how such widespread distribution came about. Seeds and viable fragments of some water-plants can be carried on the muddy feet or among the plumage of water-fowl. There is also experimental evidence that seeds of *Potamogeton*, fed to ducks and passed through their digestive system, germinated more readily than those that had not been eaten. Some doubts have been expressed as to whether these methods would be effective in the long-range dispersal of plants between continents, but it does not seem a wholly impossible idea.

Amphibians, small mammals and even insects that fly from pond to pond, such as the diving beetles, *Dytiscus* spp., and water-boatmen, *Notonecta* spp., may occasionally carry small pieces of plant for short distances. The accidental or deliberate introduction of plants imported for use in aquaria or as water garden specimens into natural waters has been responsible for the spread of some species such as Canadian pondweed, *Elodea canadensis*, and the water-fern, *Azolla filiculoides*.

4: THE SIMPLE PLANTS

Less familiar than the larger and more conspicuous plants, that were discussed in the last chapter, are the vast numbers of minute plants, many visible only with the aid of the microscope and all of comparatively simple structure, which also are present in nearly all fresh waters. These include the bacteria, the fungi or 'water-moulds', and the algae.

Although individually small, the mass effects produced when for some reason large numbers of some of them are congregated together are sufficiently striking to attract the notice of the least observant person. 'Water-bloom' on the surface of lakes, caused by floating masses of blue-green algae; the green turbidity of the water when certain forms of green algae are present in large numbers; and the almost overpowering smell emanating from ponds as a result of the activities of some forms of bacteria, are familiar examples.

The importance of these organisms in the economy of ponds, lakes and streams is in no way proportional to their size. The algae provide the first link in the food-chain, using simple chemical substances in the water and light energy to build up their own structures, thereby making available abundant food for the herbivorous animals, and, indirectly, for the carnivorous creatures. Bacteria and fungi bring about the decay and disintegration of dead animal and plant life, and by breaking down the complex organic substances contained in them, re-cycle the simpler nutrients which can be used by the plants. It has been very truly said that were it not for the bacteria and fungi, the world would soon be piled up high with dead plants and animals, and their importance is no less in fresh water than on land.

Bacteria

Traditionally, bacteria are usually included among the plants, but among simpler organisms are many that possess the characteristics of neither plants nor animals, and the bacteria could, perhaps, more appropriately be considered as belonging to a separate kingdom altogether.

Bacteria are among the smallest living organisms, single-celled (but occasionally grouped in masses or growing in filaments), with a very simple structure consisting of a central area of protoplasm surrounded by a membrane or envelope. An organized nucleus is not an obvious feature, although in recent years it has been possible to show by special techniques that some bacteria at least have nuclei. Most bacteria are incapable of moving of their own accord and depend on currents of air or water for dispersal. Some species, however, have tiny threads called *flagella* (L. *flagellum*, a whip), which project from the cell membrane, and by their vigorous lashing action propel the organisms through a liquid.

Chlorophyll is never found in bacteria, but some, including the sulphur bacteria (see p. 49), use a photosynthetic pigment, blue, purple, brown or red in colour, to convert simple chemical materials into living matter. Most of them get their food from substances which have already been built up by living organisms, and they are either *parasitic* on living animals or plants, or *saprophytic* on them when they are dead. In both cases, by breaking down the organic matter with the aid of ferments or enzymes, which they produce, they can obtain the materials they require and release other substances. Food material can be absorbed only through the cell envelope in solution.

Most bacteria require oxygen for the maintenance of their life processes, but different kinds vary in the way they obtain it. Some, called the *aerobic* bacteria, can live only in the presence of free or dissolved oxygen, which they absorb through their cell membrane. Others, the *anaerobes*, have the unique ability of existing in the absence of oxygen, deriving their energy from reduction processes in a purely chemical system (see p. 49). Many bacteria, however, can live under both aerobic and anaerobic conditions.

Three main types of single-celled bacteria occur: the spherical (*coccus*), the rod-shaped (*bacterium*) and the spiral (*spirillum*). They all multiply extremely

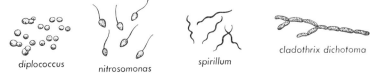

diplococcus   nitrosomonas   spirillum   cladothrix dichotoma

Fig. 13 Types of bacteria

rapidly by the simple division of the cell and the separation of the two parts, but under certain conditions some species form, within the cell, bodies called *spores*, which are capable of enduring extremes of temperature or drying-up. These spores can survive for months or even years, and, carried as wind-blown dust, are dispersed widely. Bacteria of one kind or another are found everywhere, and even when conditions preclude the existence of active organisms, spores will nearly always be present, ready to burst into activity at any suitable opportunity.

Much of the knowledge of living bacteria is now obtained by phase-contrast microscopy; details of structure are studied with the electron microscope. But if a quantity of material known to contain bacteria is introduced on to a suitable medium on which they can develop, usually for convenience contained either in a test-tube or a petri-dish, colonies of a characteristic type depending on the species will grow. One of these cultures, that of *Escherichia coli* = *Bacterium coli*, a normal inhabitant of the healthy human intestine, commonly found in water, is shown on Plate **64.**

Fresh water has its own bacterial flora, but to this are almost invariably added forms that occur on land, but have been washed in or otherwise found their way into the water. In this category come those injurious to human health, such as the typhoid bacillus, while *E. coli*, just mentioned, is always found in waters to which human sewage gains access, and an estimation of its numbers present can be used as an index of the degree of pollution of such waters.

With the parasitic types which exist only in the living tissues of animals and plants, we are not really concerned here. It is the beneficial saprophytic bacteria, those capable of leading a free-living existence and obtaining their food supply by the decomposition of dead substances, that are of such importance in fresh water. With the aid of their ferments they break down the complex compounds of dead animal and plant matter, absorbing through their cell envelope the carbon, oxygen and other substances they themselves need, and releasing less complex compounds. Several groups are engaged in this involved process of converting dead matter into simple substances which can once again be used by green plants. The first stage is the breaking down of the organic waste into carbon dioxide, water, ammonia and various ammonium compounds. Other bacteria, *Nitrosomonas*, then oxidize the ammonia to nitrites, while still more, *Nitrobacter*, oxidize these to nitrates, which then become available as food for the algae and higher plants. The mud or the mud surface is the place where this process is mainly carried out, and under certain conditions the decomposition of the organic remains by anaerobic bacteria results in the formation of sulphuretted hydrogen, with the familiar and characteristic smell of rotten eggs. 'Marsh gas' (methane), too, may be produced under these conditions, and bubbles of this gas may often be seen rising to the surface. Sometimes it becomes ignited, and the dancing flame over the water has given rise to legends of 'will-o'-the-wisp' and 'jack-o'-lantern'—wandering, elusive spirits. A group of bacteria, collectively called the *sulphur bacteria*, are capable of breaking down hydrogen sulphide, storing the sulphur from it temporarily in the form of rounded granules in their cells, and ultimately converting it into sulphates, which can be absorbed by plants. Although the results of the activities of the sulphur bacteria are, therefore, ultimately beneficial, the immediate effects are often highly inconvenient for mankind, for they may cause the corrosion of iron and steel. Underground pipes and other structures of a ferrous nature are particularly susceptible to this form of corrosion.

In ponds, large aggregations of some species of sulphur bacteria may often be found. *Lamprocystis roseopersicina*, for instance, often occurs in such quantity that the globular masses give a purplish tint to a tube of water held to the light.

When the individual bacteria are examined under the microscope, they appear iridescent, through the presence of sulphur.

If the dead remains of animals or plants are covered with water and left exposed to the light in a stoppered bottle, the development of purple sulphur bacteria will almost certainly be observed after a few weeks.

The so-called *iron bacteria*, of which there are several kinds, are filamentous bacteria which occur in stagnant, badly oxygenated waters containing iron. They obtain their energy by oxidizing the ferrous salts in the water, and precipitate the iron in the form of ferric hydroxide, akin to 'rust', in the sheath which surrounds the filament. The sheaths accumulate in time to form quite thick deposits. The reddish, rusty coating often seen covering the bottom of stagnant pools and the surface of submerged plants, stones and other objects is composed of these bacterial filaments, while some kinds form the iridescent 'oil-patches' which are present on the surface of some waters.

Crenothrix is a sufficient nuisance to have earned a number of popular names from distracted water engineers—'iron fungus' and 'water-pest' are two of the more printable of these! It is usually found as a series of minute filaments attached at one end to a solid support. Each filament is made up of a single row of rod-shaped cells enclosed in a sheath. When food material is plentiful, the individual cells burst out of the filaments, and either start new filaments or lump together into a gelatinous mass. In the past *Crenothrix* has sometimes multiplied to such an extent in reservoirs that it has turned the water red and turbid, and clogged the filters leading to the service mains.

Gallionella, another of the iron bacteria that grows in ribbons or twisted spirals, forms slimy, rusty coatings on the inside of water-pipes, and also plays a part in the formation of hard, rusty encrustations, both effects greatly reducing the internal diameter of the pipes and impeding the delivery of water. Many years ago, before the importance of iron bacteria was realized, the supply of water to Liverpool from Lake Vyrnwy dwindled to a serious extent, and it was found that the interiors of the pipes leading to the filters were almost choked with a slimy layer composed largely of *Gallionella*. Steps are now taken to minimize the effect of the iron bacteria in waterworks.

The brown, fluffy masses often seen attached to, or floating in, wells and other bodies of stagnant waters, particularly boggy, moorland pools, are formed by *Leptothrix*, the 'ochre bacterium'.

Algae

Algae are widely distributed over the face of the earth, occurring in vast numbers in the seas, in damp places on land and in fresh water. Although very many of them are of microscopic size, others are among the longest of living things, some of the brown seaweeds reaching lengths of 100 metres or more.

A microscopic alga familiar to everyone is *Desmococcus* (= *Pleurococcus*), vast numbers of which form green patches on the shady sides of trees, or on damp walls.

Many algae are composed of only one cell, some are irregular masses of cells, others form filaments made up of a single row of cells, and still others have a complicated many-celled structure. In no algae, however, do we find the differentiation of the structure into leaves, stems and roots as we do in higher plants. The reproductive processes of algae are many and varied, and sometimes quite complex.

All algae possess chlorophyll, although in some it is obscured by other pigments so that they appear blue-green, red or brown. They are, thus, like the higher plants, able to build up living matter from simple chemical substances in the presence of sunlight, and their immense numbers in fresh waters make them of great importance as primary food producers.

The differences in colour, correlated as they are with other important differences, form a convenient basis for the classification of the main groups, and these will be discussed below.

Phylum **CYANOPHYTA**
Blue-green Algae: Myxophyceae

The blue-green algae are the most primitive of all the algae, and are generally considered to be more closely related to the bacteria than to the true algae. The characteristic colour is due to a pigment, *phycocyanin*, although other pigments may also be present and give rise to the diverse colours of violet, brown or yellow seen in some species. These pigments are evenly distributed in the protoplasm, and not localized in special bodies, as in most plants. The cells are very simple in structure, without an organized nucleus, and reproduction takes place mainly by simple cell-division. No sexual method exists.

The individual cells are usually grouped together to form masses or filaments, the latter being like either strings of beads or smooth threads. *Microcystis* and *Coelosphaerium* form masses, while *Oscillatoria* is of the smooth filamentous type, and grows as a bluey-green slime on stones and other submerged objects. The filaments of *Oscillatoria* and of other related blue-greens often have curious gliding and oscillating movements. *Nostoc* forms bead-like strands which, to the naked eye, appear as round lumps of greenish-blue jelly, sometimes 25 mm across. The gelatinous covering of the filaments makes them stick to mosses. *Anabaena*, a closely related genus to *Nostoc*, is abundant in some lakes and large ponds. (Plate **9,** 1)

Rivularia appears to the naked eye as dark green, round gelatinous masses on water-plants and submerged stones. The commonest species, *R. haematites*, is found in mountain streams in limestone districts.

When conditions for growth are favourable, which is usually in late summer, some species of blue-green algae multiply greatly and, buoyed up by vacuoles of gas within the cells, float on the surface to form a scum called a water bloom. In some parts of the country the phenomenon is referred to as 'the breaking of the meres'. Sometimes algae other than blue-greens are involved in these blooms. Poisonous substances produced by the algae may kill aquatic animals, including fish, although an accompanying oxygen deficiency may be partly responsible. Cases are on record of birds and domestic animals dying after drinking waters in these conditions. The presence of blue-green algae (and also some kinds of diatoms) even in relatively small quantities can give water a definite and usually objectionable smell and make drinking water unacceptable.

Phylum CHRYSOPHYTA
Yellow-green Algae: Xanthophyceae

Two genera which appear green but are, in fact, representatives of the yellow-green algae, Xanthophyceae, are *Vaucheria* and *Botrydium*.

Vaucheria, the best known of this group, occurs as dark-green mat-like masses of filaments, rough to the touch, lying on the mud of well-aerated ponds or even on the damp earth near by. The individual filaments are long with here and there a branch, but they are not divided up into a number of cells as are the other filamentous types considered so far. Each strand is, in fact, one long tube,

Fig. 14   Vaucheria (× 30)

whence the name 'siphon algae'. Many nuclei are present along the length of the filament.

In the normal way, when food materials are plentiful and conditions generally are favourable, an asexual method of multiplication is adopted. The ends of some of the branches swell out and become divided from the rest of the filament by a wall. The protoplasm in this section then becomes dark and forms into an active spore, which eventually bursts forth and swims round for a short time by means of the flagella surrounding it, before finally settling down on the mud and giving rise to a new plant.

When conditions deteriorate, the alga resorts to the sexual mode of reproduction. Two specialized structures, one rounded and containing an egg-cell and the other in the form of a curved tube, develop on the cell-wall. From the tube, male cells are freed and swim to the egg-cell and fertilize it. The resulting resting spore can withstand the unfavourable conditions of drought or cold and, when these have passed, grows directly into a new plant.

Dotting the surface of the mud in shallow ponds, or even where the water has dried up, tiny green balls may sometimes be seen. These are specimens of *Botrydium*, another of the siphon algae. Oval in shape, they have colourless root-like structures or rhizoids, which penetrate the mud. They reproduce by means of swarm-spores, and also by the formation of resting cysts in the rhizoids. (Plate **11**, 1)

Diatoms: Bacillariophyceae

Diatoms contain brownish-green pigments which usually mask the green chlorophyll, but the group is distinguished primarily by the characteristic cell-wall.

Diatoms are among the most beautiful of aquatic organisms, and are favourite objects of study for microscopists. They are single-celled plants, all microscopic in size, whose cell-walls are largely composed of the hard, flinty substance, silica.

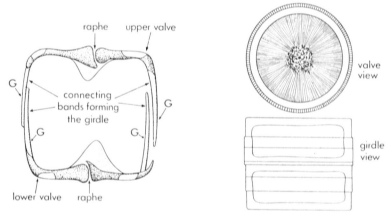

Fig. 15 Left: transverse section of the diatom *Navicula.* G = girdle.
Right: Melosira (× 500)

Each cell, or *frustule,* as it is called, is formed of two halves, one of which fits on the other like the lid of a box. The sides of the box are called the *girdle,* and the top and bottom the *valves.* Naturally, the appearance of a diatom will vary greatly, depending on whether it is viewed from the valve or girdle aspect.

The cell-walls are virtually indestructible, and on the death of the diatom they do not decay as do those of other plants, but endure as a siliceous skeleton. Vast deposits of diatomite, or Kieselguhr, consisting of the remains of marine and freshwater species of these plants exist in many parts of the world, and are used in the production of a wide range of products, from toothpastes to explosives. Some deposits in California are said to be 300 metres deep over a considerable area, and since it has been estimated that no fewer than two and a half million frustules can be contained in a volume of only one cubic centimetre, we can have little conception of the truly astronomical numbers of organisms which have gone to make up such deposits of diatomaceous earth.

The flinty skeletons are often exquisitely ornamented, particularly on the valve side, with numerous fine transverse lines, protuberances or depressions. In the living diatom, unfortunately, these are rather obscured by the colouring of the cell, the layer of mucus which is present on the outside of the frustule, and also by the low refractive index of water. By suitable treatment in acids and mounting on a microscope slide in a medium of relatively high refractive index, the full beauty of the now glass-like frustule can be seen. Even then, however, microscopical skill is needed to see the markings at their best. Slides of diatoms were formerly used as objects to test the optical efficiency of both equipment and technique. This preoccupation with the dead shells developed into something approaching a cult, whereas the study of the living organisms was sadly neglected.

Living diatoms are usually yellowish or brownish in colour, owing to the presence of pigments which mask the green chlorophyll that is also present. Some are found singly and free-floating, some are grouped together into chains or filaments, and others are attached to plants or stones by gelatinous stalks. The free-floating kinds are particularly abundant in the plankton of lakes and reservoirs, and their presence in such large numbers may be a cause of concern to water engineers. Careful examination of the plants or the surface of the mud in ponds and streams will rarely fail to reveal large numbers of diatoms, and in streams a brownish scum or slimy covering on the stones and rocks will often be found to be composed of large numbers of the stalked forms. To see the individual diatoms, however, a microscope will be needed.

Diatoms reproduce principally by cell-division. The valves become pushed apart by the increasing contents of the cell, and the latter divides into two, leaving each of the new cells with one of the original valves. A new valve then forms on the naked side. Another form of reproduction also takes place at rare intervals, two individuals merging together to form an *auxospore*, or even two auxospores, which develop into cells like the parent diatoms, but are larger.

Under favourable conditions of food supply and light, diatoms reproduce at a rapid rate, and by their photosynthesis not only augment the oxygen supply in the water, but also build up large quantities of food material which accumulate in the form, not of starch as in most plants, but of fat or oil. In spite of their indigestible covering, diatoms are eaten by many of the smaller aquatic animals.

Diatoms are broadly divided into two groups: Centrales, with valves usually circular, and Pennales, with elongated, often boat-shaped valves, which usually have a line corresponding to an opening in the cell-walls called a *raphe* running along their length. These diatoms are capable of moving along in the water, either by a gliding motion or in a series of jerks. The method by which this movement is effected remains obscure, but it is associated with the production of a film of mucilage surrounding the diatom.

Only a small proportion of diatoms are included in the Centrales, and most of those are marine. Of the freshwater forms the various species of *Melosira* are probably the commonest, occurring in ponds and ditches. The valves are circular in shape, and bear very fine markings. They are usually found joined together in short filaments. *Cyclotella* is also disc-shaped, the valves consisting of two concentric circles, the outer one bearing pronounced markings. They are rarely found in filaments, but two may be paired together.

The majority of diatoms are included in the Pennales, and only a few can be mentioned here. The rectangular frustules of *Tabellaria* are usually joined

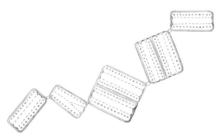

Fig. 16   Zigzag chain of *Diatoma* (× 500)

together corner to corner, in zigzag colonies, although in *T. flocculosa* var. *asterionelloides*, they are joined into a star shape similar to *Asterionella* (below). In *Meridion* the individual diatoms are wedge-shaped, but they are usually found united by their long sides to make fan-shaped, circular or spiral formations on stones or plants. The rectangular frustules of *Diatoma*, which bear pronounced transverse lines, form zigzag chains which are attached to algae. It is one of the commonest diatoms in ponds and ditches. *Fragilaria* (Plate **9, 4**) is similar in form to *Diatoma*, but the valves do not bear such pronounced markings.

Asterionella has long, narrow frustules which join at their bases and radiate outwards like the spokes of a wheel, a circular space being left at the centre of the group. The seven or eight frustules to make a complete star are not always present, and smaller groups and even single frustules are often found. The last two diatoms occur in unbelievable numbers, floating free in some lakes and reservoirs, particularly in spring. It has been estimated, for instance, that in one London reservoir alone the dry weight of *Fragilaria* present was 110 tonnes, and a tonne of the diatom was removed daily from the water by the filters. *Asterionella* sometimes gives a yellowish colour and a decidedly 'fishy' odour to the water. An oval-shaped diatom with fine markings on the valves, found abundantly on the leaves and stems of plants, is *Cocconeis*.

The family Naviculaceae contains a large number of genera, many of which are boatshaped, and have pronounced markings on the valves. Some are very common in still waters, and they include perhaps the most beautiful of all diatoms.

Fig. 17 Gomphonema (× 100)

Of the stalked diatoms, one of the best known is *Gomphonema*. The wedge-shaped frustules, borne on the ends of the branches of a fairly long stem, form a brown slimy covering on rocks or algae. *Cymbella* (*Cocconema*) is another common stalked form.

Phylum **CHLOROPHYTA**
Green Algae

This large group of the algae is one of the greatest interest to the student of freshwater life, although the very diverse forms contained in it are not confined to such a habitat, being widely distributed on land and in the sea.

In green algae the chlorophyll is not masked by other pigments, as in the other groups, so that they appear clear green in colour. The food substance formed as a result of their photosynthesis is usually starch, which accumulates round special bodies called *pyrenoids* embedded in the chloroplasts; in some kinds an oily substance is formed instead of starch. The cell-walls of green algae contain cellulose just as in higher plants.

Chlamydomonas is one of the simplest of the green algae, and is plentiful in ponds and stagnant ditches almost everywhere. A medium power of the microscope is required to observe the details of its structure, when it is seen that it consists of a single pear-shaped cell, bearing two flagella at the front end; by

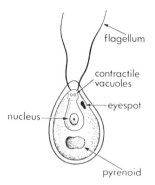

Fig. 18 Chlamydomonas (× 900)

waving these it is drawn through the water. A red eye-spot is situated near the flagella, and two clear areas will be seen near by. These enlarge and contract, and are therefore called contractile vacuoles, their purpose being the removal of waste products or excess salts from the cell. (Plate **12,** I)

Chlamydomonas multiplies with great rapidity, and in the course of a day or two its immense numbers may turn the water green. In the normal method of non-sexual reproduction, the living matter within the cell divides into a number of separate cells, varying from two to eight. These acquire two flagella, and escape from the parent cell as free-swimming spores or zoospores. At times a sexual method of reproduction is adopted. A large number of tiny bodies resembling zoospores, but smaller, are formed within the original cell. They escape, but do not give rise to new individuals directly. Instead they come together in pairs, fuse and form resting spores, covered with a tough resistant cell-wall. These are called *zygotes,* and under favourable conditions they divide up internally into a number of zoospores, each of which grows into a new *Chlamydomonas.*

Under certain circumstances some species of *Chlamydomonas* lose their flagella, and the cells, enveloped in gelatinous walls, divide and so form considerable colonies, the so-called palmella stage. When normal conditions return, the cells return to the free-swimming state.

It is significant that in higher algae reproduction often takes place by the production of free-swimming cells which, after a short journey, settle down to become new plants, and this is one of the reasons why it has been assumed that *Chlamydomonas* may be regarded as a starting-point of the evolution of the other green algae, and through them of the higher plants.

If we imagine a large number of specimens of *Chlamydomonas* all joined together in a common gelatinous envelope forming the periphery of a globe, we shall have a good idea of the structure of that favourite object of the pond-hunter, *Volvox.* The beautiful green globes (Plates **10, 12**), about the size of a pin-head, and thus large enough to be seen by the naked eye, may be so numerous in a pond or lake as to make the water green. When observed alive under a low-power microscope, particularly under dark-ground illumination, they present a singularly beautiful spectacle, as they move slowly along, rotating as they go. Up to 20,000 cells, each bearing two flagella, may be present in the outer layer of the globe of a single *Volvox globator*, but they are all connected by strands of protoplasm so that the organism may be considered a single individual.

Within the hollow sphere smaller, darker, spherical bodies of various sizes may usually be seen. Some are daughter colonies, formed asexually, which will eventually burst through the surrounding parental envelope to start an independent existence. Others will be male and female reproductive bodies called respectively antheridia and oogonia. The plates of cells making up the antheridia break away to form spermatozoids, each with two flagella. They and the oogonia are released into the water where fertilization occurs and reddish-brown zygotes, or oospores result which, because of their tough, resistant exterior, can survive unfavourable conditions, but germinate into new colonies when circumstances become suitable again.

There are three species of *Volvox* in Britain: *Volvox globator*, *V. aureus* and *V. tertius*. The largest is *V. globator*, and the individual cells of this species, which are angular in shape, are joined by thick, easily visible protoplasmic strands. The zygote has a spiny surface. *V. aureus* is smaller than the last species, with only between 1,000 and 4,000 cells, which are rounder than those of *V. globator* and joined by finer strands. This species has a smooth-surfaced zygote. *V. tertius*, the occurrence of which in this country was only recognized about 1927, is the smallest of the three species with only about 1,000 cells, which do not appear to have any protoplasmic connections between them.

Sometimes, when a *Volvox* colony is observed under the microscope, small animals will be seen in the clear space within the globe. These are rotifers (Chapter 9), either *Ascomorphella volvocicola* or *Proales parasita*, both of which live constantly in *Volvox*, feeding on the green cells of both the adult and the daughter colonies. Simpler colonies of similar type to *Volvox* are found in *Eudorina* (Plate 12, 3), which usually has thirty-two cells unconnected by strands of protoplasm; *Pandorina* (Plate 12, 2), with a compact and almost solid colony of sixteen cells connected by very delicate strands; and *Gonium*, with a squarish colony of sixteen cells. *Pandorina* and *Gonium* are common in ponds and ditches, and *Eudorina* is abundant also in the plankton of lakes, particularly in autumn.

Another colonial form of a different type is 'water-net', *Hydrodictyon*, which may be 23–30 cm in length, the hollow, cylindrical colonies being formed of elongated cells, joined together in fives or sixes to make a many-meshed network bag, floating freely in water. (Plate 11, 4)

Turning from the large to the small, mention must be made of *Chlorella*, for although so tiny that they are unlikely to be studied in their free form by the average pond-hunter, some of the species of this alga live in the tissues of aquatic animals, including the green hydra, the flatworm, *Dalyellia viridis*, freshwater sponges and the protozoan, *Paramecium bursaria*, giving them their characteristic green appearance. The relationship is apparently of mutual value, for the algae derive food material in the form of ammonia, nitrates, phosphates and carbon dioxide from the animals' life processes. In return, the animals obtain oxygen from the photosynthesis of the algae, which may be of value when the oxygen concentration of the water is low. There is evidence also that the nutrition of the animals is helped, in some cases, by the transference of organic matter to them, or by their digestion of algae in their tissues, which are either killed or have died naturally.

The species of green algae which have been mentioned so far have been either single cells or a number of cells joined together in some way to form a colony. A higher stage of development is found in what are called the filamentous algae, where the cells are arranged end to end to form a thread or filament, each cell being divided from its neighbour by a partition or *septum*. Such algae can reproduce by a purely vegetative means—that is, without the production of specialized reproductive cells—the normal cells dividing in one direction and increasing the length of the filament.

One of the simplest of the filamentous algae is *Ulothrix*, which is commonly seen in bright-green masses in rivers and streams. Other species also occur in stagnant waters, including stone drinking-troughs and the like. They are particularly abundant in spring, and seem to thrive best in cooler conditions.

It was mentioned earlier in this chapter that the reproduction of algae was sometimes complex, and this statement is well borne out by the example of such an apparently simple alga as *Ulothrix*, which employs no fewer than four methods. In the usual vegetative reproduction, the individual cells divide and increase the length of the filament. These filaments are fragile, and small portions which break off soon extend themselves in this way.

Asexual reproduction is effected by zoospores, each bearing four flagella, which may be formed in any cell of the filament. These escape through the outer cell-wall and give rise to new filaments. In the sexual mode of reproduction which takes place in early summer, smaller spores bearing only two flagella are formed in the parent cells. When these burst forth they fuse together and form a *zygote*, which itself swims round for a time with the aid of its four flagella and then settles down as a *zygospore* capable of surviving the hot weather. In autumn the zygospore divides up into zoospores like those produced asexually, and these give rise to new *Ulothrix* filaments. *Ulothrix* has still another way of reproducing for, if in danger of drying-up, the filaments divide up into a number of tiny, round green spores which can survive drought.

In *Enteromorpha*, a freshwater representative of a family mainly marine, the filamentous structure has developed into two broad, flat layers of cells which have separated to form a tube. The ribbon-like strands, often up to 15 cm long, pale green in colour and usually floating, are familiar objects in slow rivers, drainage dykes and canals, and especially in brackish stretches of water. Reproduction is by both vegetative and sexual means.

In *Draparnaldia*, tufts of very delicate threads of cells arise from the main filaments, and it is from these that the free-swimming zoospores are liberated to effect either sexual or non-sexual reproduction. The plant is large enough to be seen by the naked eye, the long green strands stretching from stone to stone

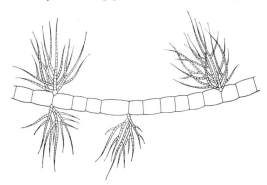

Fig. 19   Draparnaldia (× 80)

at the bottom of clear, slowly running waters or bog pools. It is a particularly beautiful object for examination under the microscope.

Cladophora, or 'blanket-weed', represents a higher stage of the filamentous structure, the branching type. The dense, dark-green masses of this large alga can be seen attached to the bottom of most slow rivers and also of lakes. In polluted rivers *Cladophora glomerata* often grows in dense masses covering a wide area, and problems may arise through de-oxygenation of the water when photosynthesis has stopped at night, and by the unpleasant smell when the alga is decaying. (Plate **11, 3, 5**)

Two species of *Cladophora*, *C. sauteri* and *C. holsatica*, form compact cushions on submerged objects and also free-floating balls, which may reach a diameter of 15 cm, but are usually much smaller, in some lakes. In parts of Japan, these 'duckweed balls', or *marimos*, are strictly protected as national monuments.

The tiny filaments of *Oedogonium* are found either free-floating or attached to submerged objects in both still and running waters, cattle drinking-troughs being favoured places. If the filaments are examined through the microscope, it will be seen that they are composed of elongated cylindrical cells, each slightly broader at the top than at the bottom, and the chloroplasts form a dense network within the cell.

Oedogonium grows by the division of cells along the length of the filament. Close examination will show, at intervals, cells with transverse rings at one end, which by careful focusing of the microscope will be seen to be caps. Division takes place at these 'cap-cells' by a somewhat complex process, and the number of caps indicates the number of times division has taken place.

This alga is also of interest on account of the unique method of sexual reproduction which occurs in some of the species. This takes place as follows: certain cells in the filament enlarge, and from these active zoospores are released, which after a time settle near the barrel-shaped cells and grow out into a very short

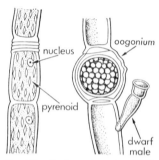

Fig. 20   *Oedogonium* (× 350)

filament. These are the co-called 'dwarf males', and from them, in due course, burst forth sperms, which penetrate to the egg-cell and fertilize it. After fertilization the egg-cell becomes surrounded with a tough, resistant wall, and in due course is set free, when the parent filament decays, giving rise, as soon as conditions are favourable, either to a new filament or to further motile forms which eventually settle down to become new plants.

Oedogonium also reproduces by asexual means.

Conjugatophyceae

From these confusing and complex methods of reproduction it is something of a relief to consider the group of green algae called conjugales, since none of them has free-swimming reproductive cells, and the sexual mode of multiplication is carried out by a process usually spoken of as *conjugation*, from which term, of course, the name of the group is derived.

The method can easily be observed in various species of the well-known alga *Spirogyra*, whose long unbranched filaments massed together in their thousands float freely in almost any body of still water, looking like green cotton-wool. Although so unattractive to the casual glance in the mass, the separate filaments are revealed in all their beauty under the microscope. The strands are then seen to be made up of single rows of cylindrical cells joined end to end, each cell having green bands—the chloroplasts containing the chlorophyll—running spirally along its length. Starch grains, the product of photosynthesis, can usually be well seen in this plant, surrounding the little bodies (pyrenoids) scattered about on the spiral bands. The filaments of *Spirogyra* are covered with a layer of mucilage, which gives the alga the characteristic slimy feeling well known to anyone who has tried to lift a mass out of the water.

Small pieces of *Spirogyra* can and do multiply vegetatively by the cells dividing and growing into long filaments. Any of the separate cells in a filament is apparently able to divide into two, and thus the filaments increase in length. It is in June or July that the sexual process or conjugation can best be observed. Two opposite cells in adjoining filaments develop swellings, which eventually meet and fuse together. The contents of one cell pass into the other, and after the fusion of the nuclei the two masses of protoplasm form an oval zygospore with a tough cell-wall. Eventually the original cell-walls disintegrate, and the zygospore is freed, developing into a new *Spirogyra* cell when conditions are favourable. (Plate **11, 2**)

The formation of a kind of zygospore has at times been observed in *Spirogyra* cells which have not conjugated, the individual cells having shrunk somewhat and developed a hard covering. These bodies are called *azygospores*.

On sunny days, among the masses of *Spirogyra* will often be seen bubbles of gas, sufficient evidence of the oxygenating value to the pond of these simple

Fig. 21   Zygnema (× 450)

plants. They are, however, not an unmixed blessing, for their dense blanket of vegetation sometimes prevents the growth of higher plants, and for this reason they are particularly disliked by owners of watercress beds.

Zygnema, a relative of *Spirogyra*, is found in similar places, but may be distinguished from it by the chloroplasts, which, instead of being spiral bands are in the form of two star-shaped bodies.

Certainly the most beautiful members of the conjugales, if not of all green algae, are the desmids (Plate **10**), roughly of the same size and having the same diversity of form as diatoms. Their lovely emerald tints and their appearance of being divided into two halves, or semi-cells, by a line or constriction (the isthmus) across the middle are, however, characters which will serve to distinguish them.

They are exclusively freshwater plants, single-celled and free-floating, although some are able to adhere to larger plants by the coating of mucilage which they possess. They are capable of moving slowly of their own volition. In sunlight they are drawn to the surface layers of the water, and if a pure gathering is needed, this is the best opportunity to obtain it. Otherwise desmids are usually found on the surface of the mud or on water-plants, and may be present in such numbers as to form a green film. Peaty pools are particularly favoured spots for them.

The cell of a desmid has a thin but firm wall, from which spines or other projections often arise. The cell contents include granular chloroplasts containing the chlorophyll. The product of photosynthesis, starch, is also present within the cell. The normal vegetative reproduction of desmids is effected by the cell dividing at the point where it narrows. In the sexual process of conjugation which occurs at intervals, two individuals merge, the cells becoming surrounded by a layer of mucilage. The cell-walls then break open and the contents unite to form a zygospore, from which in due course one or more new desmids arise.

The zygospores are very characteristic in appearance, usually being globular in shape and bearing a large number of spines hedgehog-fashion.

The species of desmids are legion, and five bulky volumes were needed by W. and G. S. West in their *Monograph of the British Desmidiaceae* to describe them. Here we can mention only four of the genera most likely to be encountered.

The many species of *Closterium*, perhaps the commonest and most widely distributed, are crescent-shaped. They may be found floating in the water or among other algae in almost any pond. The cells of *Micrasterias* are flattened and oval, with the two semi-cells divided by deep constrictions. The edges of the cells all round are indented or serrated.

Cosmarium, at first sight, resembles *Micrasterias* with the edges not so deeply indented. If seen end on, however, it will be apparent that it is oval in shape, with considerable thickness and not flattened as is *Micrasterias*.

Some kinds, such as *Hyalotheca*, are in the form of a long cylindrical thread, made up of many individual disc-shaped cells lying side by side and enclosed in a mucilage envelope. Some of the species of this genus are very abundant.

Phylum **RHODOPHYTA**
Red Algae

The red algae are mainly marine, but there are about a dozen freshwater genera which, unlike those found in the sea, are not red in colour. One of these which is frequently encountered is *Batrachospermum*, often called the 'frog-spawn alga' or 'bead-moss'. (Plate **9**, 6) Although the colour varies a great deal, it is usually greenish or brown, and the plants, large enough to see without a lens, appear as tiny strings of beads with smaller strings branching off them, bearing in the mass some resemblance to frog-spawn. The strings are attached by one end, usually to stones in running water, often in shaded situations, such as under bridges. Other species of *Batrachospermum* grow in still water, often on submerged twigs and sometimes even on the shells of living snails; large green plants may be observed in mountain tarns.

Under the lens, the 'beads' are seen to be whorls of fine filaments, and among them are often conspicuous dense, spherical bodies. These result from the process of sexual reproduction and consist of short filaments bearing terminal spores which, when released, give rise eventually to fresh plants.

Batrachospermum also reproduces asexually by *monospores*. In common with all red algae, there are no motile reproductive cells.

Stoneworts

The stoneworts are a small group of aquatic non-flowering plants which, although showing features of considerable botanical interest, have received but little attention except from a few specialists. Since they usually attain a height of about 30 cm, and often show a much-branched habit, they may easily be mistaken, at a casual glance, for higher plants. They are generally regarded, however, as a specialized group of the algae, although their peculiarities, especially their remarkable reproductive organs, have led some authorities to consider them as forming a separate major division of the Plant Kingdom—the *Charophyta*.

They grow entirely submerged, often in dense masses by themselves, in still or slowly running fresh water, which may be clean or slightly brackish. A few may occur at depths below those at which higher plants thrive. The common name refers to the hard, chalky deposit which sometimes covers almost the whole plant. This is, perhaps, more often seen on the species of *Chara*, which are commoner in alkaline waters than those of *Nitella*, the other common genus. When growing in profusion, some species give off a rather unpleasant smell of sulphuretted hydrogen.

The thallus or plant-body of a stonewort consists of bare lengths of stem separated at intervals by regions called nodes at which arise long branches similar to the main stem, and also whorls of shorter structures called branchlets, upon which are borne the complex reproductive organs. Stoneworts have no asexual reproductive stage, although fresh growth may arise from an old node.

In most species both male and female parts are carried on the same plant, in close proximity to each other (Plate 10, 5), but in a few species the plants are separately sexed. The female *oogonium* is oval, with an outer envelope of five spiral cells and an inner nut-like body comprising the single egg-cell. The outer membrane of the ripe 'nut' is often characteristically patterned.

The male *antheridium*, which is spherical, has an outer covering of eight triangular, beautifully sculptured shield-plates, which on maturity break apart, usually in the morning, and release a large number of filaments, each of which contains numerous male cells or *antherozoids*. These make their way to the female cell and fertilize it, to produce a thick-walled spore which falls from the parent plant and in due course gives rise to a new plant.

Stoneworts have no roots in the strict botanical sense, but root-like structures, or rhizoids, penetrate the mud and serve to anchor the plants. From these rhizoids, and also sometimes from the lower stem nodes, whitish swellings called bulbils are produced, which are accumulated reserves of starch and from which new plants arise by vegetative means.

Each internode of a stonewort consists of a single cell (although in *Chara* this is usually covered by an outer, protective, many-celled cortex) that may sometimes be several centimetres long. The plants, when not heavily encrusted, are so translucent as to make it particularly easy to observe under a microscope the phenomenon known as cyclosis, or the rotation of protoplasmic particles in these cells.

The stoneworts are classified into two sub-groups: the Nitelleae, the members of which have forked branchlets and no cortex; and the Chareae, with unforked branchlets and usually a cortex. In the former division are included the genera *Nitella* and *Tolypella*, while the latter comprises the genera *Nitellopsis* and *Lamprothamnium* (each with a single species) and *Chara*. In all, there are over thirty British species of stoneworts and some twenty named varieties. (Plate **13**)

Fungi: the Water-Moulds

Like the bacteria, fungi do not possess chlorophyll, and to obtain their food they are either parasitic on living plants or animals, or saprophytic on dead ones. Both types exist in fresh water. Perhaps the best known, and the most obvious in their effects, are the Saprolegniales. Specimens of this group may usually be obtained by 'baiting' pond water with small pieces of dead animal or plant tissue. Hemp seeds (obtainable from pet shops) make particularly fruitful baits. They should be boiled vigorously, or split before use, and then left in shallow dishes in contact with the water collections for six to seven days, by which time white tufts of water-moulds growing on the seeds will be visible to the naked eye. If these tufts are examined under a lens, or preferably a microscope, they will be seen to consist of numerous individual filaments, the tips of which accumulate dark contents and become cut off by cross walls to form *sporangia* or *oogonia*. Numerous free-swimming spores are released from the sporangia and these can settle down and grow if a fresh substratum is presented to them. The oogonia contain resistant reproductive bodies which can survive adverse environmental conditions. In addition to the filaments which bear the reproductive organs, other less conspicuous ones penetrate the storage tissues of the hemp seeds and produce enzymes which break down the complex food material into an easily assimilated form.

Most of the Saprolegniales are purely saprophytic and perform a useful service in nature by breaking down dead plant and animal remains. Others are facultative parasites, however, and these are responsible for the fungus diseases which affect freshwater fish. They are probably the biggest scourge in trout farms or trout waters generally. The fungus occurs at first as small whitish patches on any part of the fish, particularly where there has been some wound or bruise. It spreads until quite large areas are covered, and eventually may cause the death of the fish. Goldfish and other fish kept in garden pools or aquaria are often infested with this fungus, usually when their skin has been damaged or bruised. Malachite green in very low concentration (two parts per million or less) inhibits the fungus and this substance is sometimes added to the water in hatcheries since developing fish eggs are particularly susceptible to attack.

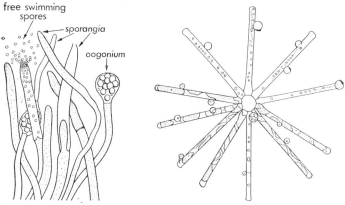

Fig. 22   Left: Saprolegnia filaments showing oogonia and sporangia, one of which is releasing free-swimming spores (× 60). *Right:* Asterionella colony heavily parasitized by a chytrid, *Rhizophidium planktonicum* (× 300)

A more primitive group of aquatic fungi is the Chytridiales, some members of which are obligate parasites and others saprophytes. The parasitic types are found mainly on algae and microscopical animals, and often result in their death. They sometimes appear suddenly in great numbers in some waters, and recent investigations indicate that they may have a considerable effect in reducing the numbers of algae in lakes. Chytrids reproduce both by sexual and non-sexual means. In the latter, free-swimming spores are formed which seem able to make their way to suitable hosts, where they develop into new plants.

5: THE PROTOZOA

The Protozoa (Gr. *protos*, first; *zöon*, an animal) are the animals with the most simple bodily structure but they show such variations among themselves that they are usually considered to constitute a separate sub-kingdom. It is customary to call them single-celled, but non-cellular is perhaps a better description, since although they are *structurally* simple, they exhibit considerably more intricacy than ordinary single cells, and are complete living creatures with the capacity of actively taking in food, moving about, reproducing and of ridding themselves of waste products.

All Protozoa are very small, some being exceedingly minute, and a microscope is needed to study them, but size alone is not a distinguishing feature, for a number of many-celled animals, which are considered in a later chapter, are smaller than some of the Protozoa. Most are free-moving, but some attach themselves by stalks to solid objects, while others form loose colonies, these latter pointing the way by which the many-celled animals or Metazoa (Gr. *meta*, later in time), which constitute the rest of the Animal Kingdom, have evolved.

More kinds of Protozoa live in the sea than in fresh water, and the vast deposits of chalky shells of foraminifera and flinty shells of radiolarians that lie on the sea-bed, in some parts of the world, bear witness to their abundance in earlier times. Others live on land, in damp soil, and many are found inside the bodies of other animals.

Most Protozoa multiply by dividing into one or more individuals, and this form of asexual reproduction can go on very quickly under suitable conditions. In some Protozoa, two individuals come together in conjugation at certain times, and there is an exchange of nuclear material, after which they separate. This mode of reproduction seems necessary occasionally to maintain vigour and genetic variability.

The Protozoa and other apparently simple organisms which were considered to be related to them were formerly classified in one heterogeneous group, the *Infusoria*, because they appear in a 'hay infusion'; that is, water in which a small quantity of hay or other dried vegetable matter is left for a few days.

They nearly all possess the ability to withstand drying-up by surrounding themselves with a resistant covering, and their desiccated forms, left perhaps on aquatic vegetation by the lowering of the water level, or maybe stranded on the mud at the edge of a pond, are blown about in the air in such numbers that almost any plant will be covered with them. When an 'infusion' is made, they burst into activity again as if nothing had happened.

Class *MASTIGOPHORA* (= *FLAGELLATA*)

In studying the most primitive forms of life, it is not always easy to tell what is animal and what is plant, and this group at least can be claimed with some justification by both botanists and zoologists. The flagellates are, however, included here among the Protozoa. They all bear hair-like processes called flagella at the front end of their bodies, and these by their lashing action pull the organism through the water.

The species of *Euglena* are perhaps the best-known flagellates and, although they are of a somewhat advanced form, they will serve as an example of the class. *Euglena* is found often in great abundance in ponds and other small bodies of stagnant water, particularly those rich in organic matter, such as farmyard ponds, sometimes forming a green scum or colouring the water green. Under

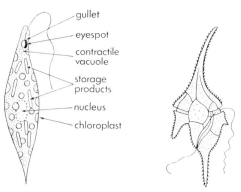

gullet
eyespot
contractile vacuole
storage products
nucleus
chloroplast

Fig. 23   Left: Euglena (× 250)
Right: Ceratium (× 150)

the microscope there will be seen within the protoplasm numerous chromatophores, usually containing the chlorophyll which enables the organism, like a plant, to utilize sunlight to build up starch-like products. (Plate **12**, **10**)

Under conditions where chlorophyll is not needed—as for instance in a dark place, or in water rich in food material—some species of *Euglena* become colourless and absorb nutrient solutions through the cell surface, thus behaving like animals. In spite of the name 'gullet' given to the channel at the front of the body, *Euglena* does not ingest solid food.

The *flagellum* is not always easy to see in a living specimen, but its presence may be inferred by the jerky movements of the body. A short and possibly vestigial flagellum is located near the base of the normal one, but it has no function in locomotion. Near the main flagellum will be seen a small orange-coloured area, the so-called eye-spot, a point sensitive to light. A small, clear area which fluctuates in size is the contractile vacuole in which is collected the surplus water in the cell and perhaps also waste products.

Dinoflagellata possess two *flagella*. An example is *Ceratium*, which is exceedingly abundant in some lakes, floating freely in the water. The cell-wall is made up of a series of plates with characteristic long horns or prongs.

Class *RHIZOPODA* (= *SARCODINA*)

The next major group of Protozoa is Rhizopoda (Gr. *rhiza*, a root; *podos*, a foot), of which the best-known member is *Amoeba*. For a long time this interesting creature was regarded as the simplest form of living matter, and is still often referred to as such in popular scientific writings. To the earlier naturalists it seemed to symbolize the 'primeval blob of jelly', which they fondly imagined represented the first stage in the evolution of living matter on the earth, but the

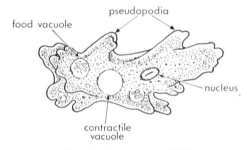

Fig. 24   Amoeba (× 150)

discovery of bacteria and other lowly forms have made *Amoeba* seem, by comparison, to be highly developed. This simple, shapeless mass shows all the essential functions of a living animal—movement, feeding, excretion, respiration, reproduction and sensitivity to its surroundings.

Several species of *Amoeba* are found in ponds and ditches, gliding over the surface of the mud. They are best obtained for study by skimming the mud with a wide-mouthed tube or bottle and by leaving the water to settle for a day or so. If amoebae are present, they will then be seen on the glass sides as minute greyish specks just visible to the naked eye when the bottle is held against a dark background.

The first examination of an *Amoeba* under the microscope is apt to be disappointing. Instead of the irregularly shaped animals with well-defined limbs, clearly marked nucleus and vacuoles, which everyone is by now accustomed to seeing in books and magazines, there will merely be a shapeless, jelly-like mass

on the stage, in which little of the structure of the organism can be made out.
The examples from which illustrations are taken are, of course, prepared specimens which have been stained with dyes to reveal the various parts. (Plate **14, 1**)

The body of an *Amoeba* consists of two layers of protoplasm, the internal,
granular *endoplasm* and the clear, thin layer of *ectoplasm* at its outer margin.
When the creature moves, the endoplasm flows out into long protrusions called
pseudopodia (Gr. *pseudos*, false; *podos*, a foot), and the rest of the protoplasm
follows in the same direction. Should food material be encountered—for example, diatoms, green algae or bacteria—the pseudopodia flow round it and,
surrounded by a film of water, it becomes engulfed in the body, where digestive
ferments reduce it into an assimilable form. The space in which the food is enclosed is called a food vacuole, and there may be any number of these food vacuoles within the endoplasm according to the abundance of food available to the
creature. Any indigestible portion of the food gradually finds its way to the surface of the body, and is left behind as the creature moves on.

Other vacuoles which are constantly expanding and contracting are also
present within the body-mass; these *contractile* vacuoles are for the purpose of
excreting surplus water and possibly small amounts of waste products. The
elimination of most of the waste products is, like respiration, carried out through
the body surface.

Amoeba reproduces by simple division. The creature first of all becomes more
or less spherical with no pseudopodia protruding. A 'waist' then develops which
gradually nips the body into two identical but smaller amoebae, which move
away from each other to lead their own lives. Under unfavourable conditions

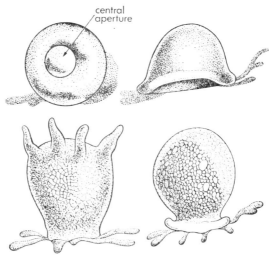

Fig. 25   Above: Arcella (× 350)
Below: Difflugia corona and D. urceolata (× 75)

such as the drying-up of the pond, *Amoeba* may surround itself with a tough protective coat or cyst in which state it remains until conditions improve. Within the cyst, the body sometimes divides into many parts, and when conditions return to normal these emerge as new individuals.

Amoeba is merely a mass of naked protoplasm, but there are other groups of rhizopods which protect themselves with a covering. *Arcella* secretes a smooth shell made of chitin, the substance which forms the outer covering of insects. The animal is shaped rather like a basin with the convex surface uppermost, and is found in the mud surface of ponds, particularly those rich in organic matter.

Difflugia, which is found in similar habitats, has a pear-shaped shell which is composed of sand grains and debris attached to a thin, secreted layer. Both these creatures extend pseudopodia out of their shells and move along like *Amoeba*.

Class *ACTINOPODA*

These animals have pseudopodia which are long, slender and of more or less constant shape. They stand out from the spherical body of the animal like spines, and their resemblance to the rays of the sun has given the creatures both their scientific name, heliozoans and also their popular name of 'sun-animalcule'.

Fig. 26   Actinophrys sol (× 500)

Actinophrys sol is perhaps the commonest species in ponds, being particularly abundant in those containing decaying vegetable matter. Its body contains many vacuoles, appearing frothy in consequence, and bears numerous 'spines' all over it. The creature rolls along by slightly contracting the pseudopodia in turn. Should minute organisms or algae be encountered, the pseudopodia bend over and surround them, and they become absorbed into a food vacuole in the body. *Actinosphaerium* is much larger than *Actinophrys*, has coarser rays and numerous nuclei, in place of the single central nucleus of the smaller species.

Class *CILIOPHORA*

The ciliates are characterized by *cilia* (L. *cilium*, eyelash), fine hair-like processes, which by their rhythmic waving action enable the creatures to move along, or to create currents in the water which will bring to them floating food particles.

They are, perhaps, the most familiar of the Protozoa and some are exceedingly abundant in most freshwater habitats. The colonial forms are particularly beautiful objects to examine under the microscope, especially if dark field illumination is used.

Most belong to the sub-class Ciliata and some are free-swimming, whereas others live attached to plants and other objects.

Order HOLOTRICHA

Of the free-swimming forms *Paramecium* (of which there are several species) is the best known. Its popular name of 'slipper animalcule' refers to its elongated body, rounded at both ends, which, with the added feature of a deep fissure on one side where the mouth is situated, closely resembles footwear in appearance. It is sufficiently large to be seen without the aid of a lens as a greyish speck swimming in the water, if the glass tube containing a culture is held against a dark background.

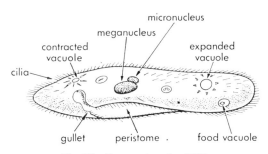

Fig. 27   Paramecium (× 350)

Unlike *Amoeba*, *Paramecium* maintains a constant shape, and has a definite outer layer of skin. Within the body there are, in addition to the usual vacuoles, two *nuclei* called respectively the *meganucleus* and the much smaller *micronucleus*, of importance in the complex process of reproduction. The whole outer surface of the body bears the cilia, and the rhythmic beating of these propels the creature quickly along. If it encounters an obstacle, it is capable of reversing the action of the cilia to move backwards. Further cilia present in the mouth aperture sweep food—bacteria and other organic particles—down into the gullet, where, enclosed in a food vacuole which moves on a definite course inside the creature, it is digested.

Beneath the skin of *Paramecium* are embedded small oval bodies called *tri-*

chocysts, the purpose of which is not quite clear. They can be discharged outside the body through pores, and in the process become long fine threads. It has been suggested that the trichocysts may serve as a form of protection for the animal.

Paramecium reproduces mainly by dividing into two, but this simple method cannot be carried out indefinitely. A time comes when the creatures lose vitality and, in order to maintain vigour, it becomes necessary for two individuals to unite in conjugation and exchange portions of their nuclei. When this is about to happen the organisms become sluggish in their movement, and eventually two come together and adhere peristome to peristome (Fig. 27). Parts of the nuclei of each pass into the other, and they then separate, both rejuvenated and each ready to divide again a number of times. In this form of reproduction there is no immediate multiplication, but it seems to be a necessary stage in the propagations of the creature and represents an early manifestation of the sexual process, although the individuals taking part are, as far as can be seen, the same in every respect.

Paramecium is very plentiful in almost any water where there are decaying leaves or other vegetable matter. A convenient spot to find it is in water lodging in a leaf-filled house gutter.

Fig. 28   Left: *Colpidium* (× 350) Right: *Loxodes* (× 300)

Paramecium caudatum is the largest species of those commonly found. *P. aurelia* is less pointed at the rear end and has two micronuclei. *P. bursaria* is easily recognized by its green colour due to the presence of symbiotic algae. (Plate **14,** 2, 3) In infusions for culturing *Paramecium* another smaller ciliate appears first. This is *Colpidium,* which is bean-shaped, but with a narrowing of the body towards the anterior end.

Loxodes is about the same size as *Paramecium,* but its body is beak-shaped at the front end and is very flexible, being bent this way and that as it moves through the water.

Order SPIROTRICHA

Members of this order have a permanently open gullet, which is provided with an undulating membrane and a wreath of cilia around the mouth.

Spirostomum is common in some ponds, and has a much longer and more slender body than the preceding animals, looking almost like a worm. Both this genus and the next are interesting in having what is called a moniliform nucleus— that is, a nucleus made up of a number of smaller nuclei in a line resembling a string of beads.

The various species of *Stentor*, although differing somewhat in form among themselves, are in the main conical or funnel-shaped, and often attached by their narrow end to a solid support. Their bodies are covered with small cilia, and there is a crown of longer ones at the wide open end. By the waving motion of the crown, currents of water are drawn towards the creature, bringing food particles to the mouth. Unwanted substances can also be wafted away by a reversal of the motion of the cilia.

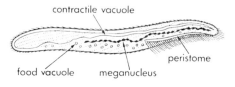

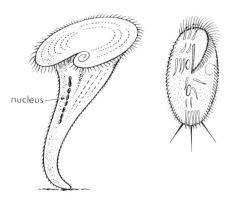

Fig. 29   Above: *Spirostomum* (× 30)
Below: *Stentor* (× 60) and *Stylonchia* (× 200)

Stentors are very common in some ponds, sometimes attached to plants, sometimes free-swimming, but when swimming they assume a rounder shape. The larger species can just be seen with the naked eye, as they are about 1 mm in length when extended, but if disturbed they contract into a little blob not easily noticed. They may be bluish or brownish in colour, with considerable variation even between individuals of the same species. Some are also green owing to the presence of symbiotic algae, although *S. lorica*, a large species that builds a case or lorica, is so coloured with its own pigments. *S. coeruleus* is the largest stentor and, as its name implies, of a cerulean blue colour.

Stylonchia has a flattened body and stiff, fused cilia which it uses to crawl over algae or other plants. *Euplotes*, another common ciliate, also has bristles for the same purpose but is smaller. It is often seen in the process of cell division. *Kerona* is an external parasite of *Hydra* (see p. 86).

Order PERITRICHA

The colonial and social forms of the ciliates are exceedingly common in most waters, attached to water-plants, dead sticks, stones or even the bodies of other aquatic creatures, the movements of which they may thereby impede greatly.

The bell animals, as the various species of *Vorticella* are popularly called, are found in dense masses at times, looking at first glance rather like a fungoid

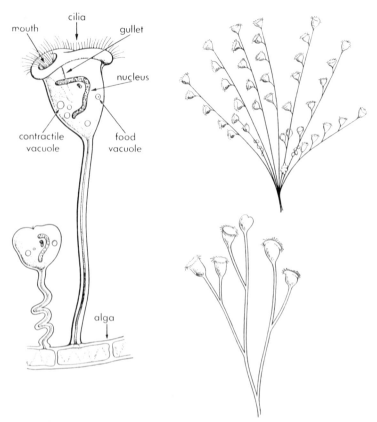

Fig. 30   Left: Vorticella (× 100) *Right: (above) Carchesium*
(× 30) : (*below*) *Campanella* (× 30)

growth on the plant or other support. *Vorticella*, however, is not truly colonial in the biological sense, since the individual animals are not connected in any way with their neighbours, each stalk being separately attached to the support. The body of the organism is like an inverted bell, and bears a crown of rapidly vibrating cilia which bring food to the mouth. When the creatures are disturbed, as for instance when the microscope stage on which they are being examined is tapped, their stalks contract suddenly into a corkscrew shape and the 'head' closes up, only to expand again in a few seconds. (Plate 14, 4)

At times the 'heads' become detached from the stalks and swim freely in the water. In at least one species, *V. campanula*, a type of swarming has been observed, all the animals from one group leaving their stalks at the same time and swimming round, held together in a film of mucus which they have secreted. After a time the swarm settles down and the individuals grow new stalks.

In *Carchesium* the bell-like animals are united, sometimes in hundreds, into a proper colony, each stalk connected to a main stem common to them all. This is in turn attached to solid objects. The main stem rarely contracts, but the individual animals contract and expand on their stalks like *Vorticella*.

Another colonial form, *Campanella* (= *Epistylis*), has all the animals united on a stiff main stem which cannot contract, and when disturbed the 'heads' merely close up and nod.

Reproduction in the vorticellids is effected by simple division of the 'heads', one of the new animals usually remaining on the stalk and the other swimming away to settle elsewhere. There is also a sexual mode with the conjugation of two separate individuals which have broken away from their stalks and become free-swimming.

Ophrydium occurs in dense, roughly circular masses, the short branching stalks bearing the individual animals being embedded in a jelly-like substance. They are usually among water-plants, and since, like other vorticellids, they may be green through the presence of zoochlorellae, they are easily mistaken for a part of a plant. I well remember my surprise on the first occasion I encountered this organism. When examining what I took to be a resting bud of a bladderwort plant, I found it was a colony of *Ophrydium*.

Fig. 31   Left: Acineta (× 600) *Right: Dendrocometes* (× 400)

Order CHONOTRICHA

This small order contains *Spirochona*, the species of which live attached to the gills of the freshwater shrimp, *Gammarus*, where they resemble tiny, slender vases.

Sub-class *SUCTORIA*

The members of the sub-class Suctoria differ from the ciliates considered so far because the cilia are only borne by a free-swimming early stage and are lost when the animals are fully developed. Most suctorians that will be found are attached to plants, but others live a free existence; some are internal parasites of other animals. *Acineta* has a rigid stalk bearing a cup-shaped body from which protrude extensible, sticky tentacles with which prey is caught. *Dendrocometes* has a lens-shaped body from which branching arms ending in tentacles arise (Fig. 31). It is found attached to the gills of *Gammarus*.

Class *SPOROZOA*

The members of this class are internal parasites of other animals. Their life-histories are very complex, often involving two or more hosts. *Plasmodium*, the organism which causes malaria in man, for instance, goes through several stages in the body of an adult anopheline mosquito, and it finally reaches the salivary glands of the insect, from where it is injected into the blood-stream of a human being when the victim is bitten. The parasites then enter into the cells of the reticulo-endothelial system in the liver, brain, spleen, etc., and go through

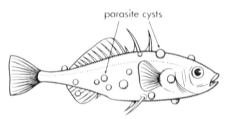

Fig. 32 Stickleback parasitized by the sporozoan *Glugea*

a cycle of development there. From this seat of infection the blood corpuscles are later invaded by small trophozites. Eventually, a stage of the parasite is sucked out of the blood of the infested person when he is bitten again by a mosquito and the cycle starts again.

Among the aquatic Sporozoa, probably the only kind likely to be encountered by the average student of pond life is *Glugea*, which attacks fish, particularly the common three-spined stickleback, *Gasterosteus aculeatus*. Occasionally these fish are found with white nodules on various parts of their body. These are caused by the presence of *Glugea*, and in due course large numbers of spores are released from the nodules into the water and infect other sticklebacks.

Lankesterella is parasitic on frogs, living in their blood-vessels. When a leech attacks a frog and sucks up its blood, the protozoan is transferred to the new host, and undergoes the rest of its life-history within it.

In this chapter it has been possible to mention only a few typical representatives of the main groups of Protozoa. The abundance of these small, and apparently insignificant, animals in a normal freshwater habitat is shown by a study carried out in 1961 by Dr Marjorie Webb of Leicester University who found more than 120 species, of which 90 were ciliates, in the mud of a small eutrophic lake in Lancashire. Such numbers indicate that they must form an important food supply for the larger animals, especially when these are young. The small size of the prey is offset by their rapid reproduction. Only recently has it been shown that some groups of Protozoa, notably in Rhizopoda, graze on planktonic algae, especially colonial green algae, to such an extent as to destroy almost the whole population in a lake within 7–14 days, and are thus of startling ecological significance.

Some Protozoa form a not inconsiderable part of the plankton of lakes, and as predators and parasites of algae may affect the populations of some of them. Other parasites, such as *Glugea* (p. 77) kill larger creatures, including fish.

Protozoa that engulf bacteria have particular significance in polluted waters and will be mentioned again in Chapter 19. In the various stages of organic pollution of a river, a natural succession of Protozoa can be seen. In a badly polluted river, where bacteria are the principal living things, only a few species of Protozoa such as *Bodo* occur. As conditions begin to improve, bacteria-eating ciliates such as *Colpidium* and *Paramecium* become abundant, followed by *Vorticella* and the colonial *Carchesium*, with algae-eaters such as *Stentor* and *Spirostomum*. Protozoa, therefore, may serve as useful indicators of stages of organic pollution.

Finally, to the biologist, the colonial forms of Protozoa are of interest as pointers to the direction from which have come the many-celled animals which will be the subjects of succeeding chapters.

6: THE SPONGES

It may come as a surprise to some readers to learn that there are in fresh water relatives of the sea sponges whose horny skeletons are used in bathrooms. There are, however, only two common species in the British Isles—the pond sponge, *Euspongilla lacustris*, and the river sponge, *Ephydatia* (= *Spongilla*) *fluviatilis*, for the phylum Porifera (L. *porus*, a pore; *fero*, to bear) is essentially marine, and only a few members of it have colonized fresh water.

The names of our two common sponges are unfortunate and misleading, for the species are not confined to the localities attributed to them. The river sponge occurs most frequently in lakes and other still waters, forming flat encrusting growths on plant stems, tree rootlets, the underside of floating logs and wooden posts in the water. The pond sponge is usually found in rather deep water in the stiller parts of rivers and canals, growing, with many finger-like branches, from the bottom or the sides of locks. A good place to search is under canal bridges. The normal colour of both species is greyish or yellowish-white, but quite often they may be green through the presence in them of the unicellular alga *Chlorella*. Seen thus, they might easily be mistaken for plant growths, and they were, in fact, regarded as such by the older naturalists.

Sponges have been claimed as a separate sub-kingdom Parazoa, intermediate between the Protozoa on the one hand and the many-celled animals, or Metazoa, on the other. They might be considered almost as colonies of Protozoa, for although their structure may consist of many thousands of cells, these are not differentiated into special organs, and they retain a considerable independence in their working. The cells of a sponge are in two layers arranged round a central cavity. If the surface of one of our freshwater sponges is examined through a lens, numerous holes or pores will be seen from which the name Porifera is derived. Some of these are small and some large. Through the smaller pores water is constantly taken in and passed through the canals which ramify throughout the creature's structure, eventually passing out again through the larger openings or oscula. By this constant circulation of water, which can be shown by placing a little carmine in the water near a sponge, food particles, oxygen and silicates (from which the sponge builds its skeleton) are brought in, and waste matter and carbon dioxide carried away. The currents are generated by the lashing action of multitudes of cells, each bearing a single flagellum, which line the inner cell-layer. These cells, which digest the food, closely resemble some of the collared flagellate Protozoa, and it is supposed that sponges evolved from colonies of such organisms.

Between the two cell-layers is a mass of jelly into which wander cells of various kinds; some are like *Amoeba* and carry the food digested by the flagellate cells round the sponge; others give rise to reproductive bodies, and still others produce the spicules which will form a supporting structure or skeleton for the sponge. The skeleton of a bath-sponge is made of *spongin*, but in the freshwater species silica is secreted in the form of needle-like *spicules*. These can be seen best by heating a piece of the sponge in strong nitric acid in a test-tube to dissolve the surrounding tissue and then examining the residue under a microscope. If the spicules are all smooth, spindle-shaped and slightly curved, the species is *Ephydatia fluviatilis*; if there are also slender spicules about a quarter the length, it is *Euspongilla lacustris*. (Plate **15**, 3)

In days gone by, freshwater sponges dried and ground to a powder were used, particularly on the Continent, for a variety of purposes. Rubbed on affected parts of the body, the powder was supposed to be a cure for rheumatism, presumably because the spicule particles caused an irritation and a local warmth. It was an official cure for 'black-eyes', and it was often prescribed as a substitute for capsicum as an embrocation for cholera. Girls in Russia used the powder as a form of rouge to redden their cheeks, and at the end of last century it was sold in England, under the name of 'Russian fleas', to mischievous small boys, who would place small quantities under their playmates' collars to cause violent itching!

After the brief description which has been given of the mode of life of freshwater sponges, it will not be difficult to understand which locations will best suit them. The water must be clear for much suspended matter would choke the pores and canals. It must contain enough silicates for the creatures to abstract sufficient for their supporting spicules. Finally, as sponges do not seem to like 'hard' water, it must not contain much lime.

Sponges increase by growing buds, which usually do not separate off but merely increase the size of the animal. They also produce *gemmules*, which are clumps of cells produced internally and covered with a tough, resistant layer. These can often be seen towards the end of summer as dark-brown spots about the size of a pin-head dotted about the body of a sponge. When the main sponge body dies down in winter, these fall to the bottom of the water and remain dormant until spring, when they give rise to small sponges.

During the summer months sponges also reproduce sexually, both male sperms and eggs being developed (in different individuals) in the jelly-like layer between the cell-layers. The sperms are set free into the water, and are drawn by the current into another sponge, where they fertilize its eggs. The spherical larvae which hatch from the eggs, at any time from May to October, have cilia all over them, and when they are liberated through the large pore of the parent sponge, they swim round for about a day before settling down on some solid object and developing into a tiny sponge. During this development a very interesting change takes place, for the creature turns inside out. The outside cells bearing the cilia migrate inside and become the current-producing flagellated cells of the inner layer, while the inner cells become the outer.

Sponges, then, are by no means the uninteresting objects they appear at first sight. Although made up of many cells, they bear little resemblance to the rest of the multicellular animals, for they have no organs, no mouth, no stomach or anything really comparable to a food canal, and no nervous system. Their two layers of cells are not like the ectoderm and endoderm of other animals. Their affinity is more with the Protozoa, and it is not surprising that they are sometimes placed in a separate sub-kingdom by themselves, the Parazoa (Gr. *para* beside, near; *zöon*, animal).

Their part in the economy of fresh water is small, and perhaps their most important role is that of providing shelter for a multitude of creatures. In particular the larvae of spongilla flies, *Sisyra* (see p. 183), live on the surface of, or embedded in, the sponges, extracting from them their food. To man, sponges sometimes become nuisances by their prolific growth in water-pipes, where they may block the pipes and give unpleasant tastes to drinking water. The efficient filtration practised in modern waterworks has minimized this trouble.

If sponges are kept for study, the mistake should not be made of thinking that as they do not move they can be placed in any small receptacle that is handy. They pass through their bodies, in the course of a day, considerable quantities of water, and rapidly die if they are deprived of a plentiful supply.

7 : HYDRA AND ITS RELATIVES

The phylum Coelenterata (Gr. *koilos*, hollow; *enteron*, intestine) includes such well-known marine animals as corals, sea-anemones and jellyfish, but the only common freshwater representatives are the species of hydra. These creatures, which when fully extended may be as much as 25 mm long, have a hollow tube-like body bearing at the top a crown of four to ten long tentacles. Hydras are usually plentiful in weedy ponds and stagnant ditches, attached by their base to water-plants, particularly the undersides of floating leaves as varied in size as those of water-lilies and duckweed. On lake shores, and in running waters, the undersides of submerged stones are often rewarding places to search for hydras.

Although they are visible to the naked eye when extended, they contract to minute, almost invisible, blobs when disturbed, and the best way to obtain specimens for study is to place a handful of duckweed or a few lily leaves in a glass vessel near a window. After a day or so, if any hydra are present, some will almost certainly have attached themselves to the glass, and they can be picked off with a pipette. The easiest method of doing this is to dislodge them gently with the tip of the tube, and then suck them up quickly.

Some uncertainty exists as to the number of species of hydra that are to be found in this country, but three are widespread: the green hydra, *Hydra viridissima* (= *Chlorohydra viridissima*); the brown hydra, *Hydra oligactis* = *Hydra fusca* = *Pelmatohydra oligactis*; and the slender hydra, *Hydra attenuata* = *Hydra vulgaris attenuata*. There can be no doubt about the identity of the first of these species, for its brilliant, emerald-green colouring, due to the presence in the inner cells of its body of minute algae, *Chlorella*, is a unique feature. Its tentacles are short and never as long as the body, which may be 20 mm in length. The brown hydra is recognized by the slender stalk at the base of the body column, but this feature is sometimes absent, and in the laboratory stalked forms have given rise to stalkless individuals. The tentacles are long, and when fully extended may be four or five times as long as the body, which may exceed 25 mm. The slender hydra, which was previously considered to be merely a variety of the common hydra, *H. vulgaris* = *H. grisea*, found on the Continent, is smaller than the two previous species, the body column being only about 15 mm long when extended. Its tentacles range in size from less than the body length to nine times longer. It has no stalk, the column being roughly the same thickness all the way down. Other species believed to occur in Britain are *Hydra circum-cincta* and *H. graysoni*.

In structure, the body of a hydra is rather like a fur-lined gauntlet, except that it is cylindrical and not flattened like a glove, and the opening is at the wrong end. The outer skin of the gauntlet and the fur lining would be comparable to the outer and inner cell-layers—the *ectoderm* and *endoderm* respectively—for in the hydras, as in the sponges, the body-wall is made up of two layers of cells, which are separated by a non-cellular jelly-like substance.

The hand and wrist portion of the gauntlet would represent the large space in the middle of the body, which serves as a gut-cavity. There is no body-cavity in a hydra separate from the gut-cavity, as in higher animals. Finally, the fingers of the gauntlet, although perhaps insufficient in number, would

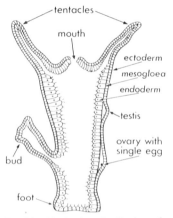

Fig. 33 Hydra: longitudinal section

resemble the hollow tentacles, which all the time are waving about in search of food.

The tentacles surround a conical area, at the apex of which is the only opening to the body, and through which all food is taken in and waste materials expelled. The skin-cells of the ectoderm are roughly conical in shape, and their narrow inner ends are prolonged into slender processes, which act as muscles. Similar muscle processes occur also in the inner layer, and their joint action enables the creature to perform its spectacular changes of shape and size, now contracting its body to a mere round blob, now expanding it into an almost unbelievably long tube. A primitive nervous system connects the muscles, so that the effect of a stimulus acting on any part of the body is quickly transmitted all over and brings about a general contraction.

Between the skin-cells there are smaller 'packing cells', and from some of these, particularly on the tentacles and upper part of the body column, develop what are called *cnidoblasts* (Gr. *knide*, nettle; *blastos*, a bud, shoot). It is in these cnido-blasts that are found the remarkable harpoon-structures, the *nematocysts* (Gr. *nemat-*, a thread; *kystos*, a bladder, pouch), with some of which the hydra captures and paralyses its prey.

The nematocysts, of which there are four types in each animal, vary somewhat in their structure in each species and can be used as a means of identification. They are hollow sacs, roughly pear-shaped, and the outer, narrow end is tucked

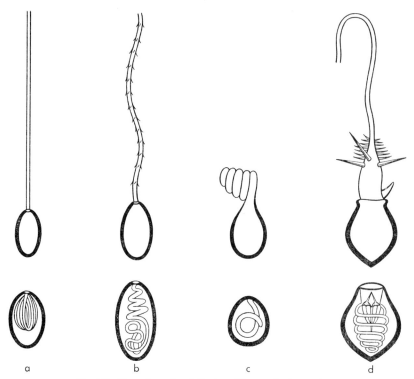

Fig. 34   Nematocysts of *Hydra* : *Above:* discharged
Below: undischarged. a. and b. glutinants, c. volvants, d. penetrants

in to form a long, hollow thread normally coiled up inside the sac. The effect is similar to turning the finger of a glove inside out. The nematocysts contain liquids, the nature of which differs between the four types. When the end of the nematocyst is stimulated, for example by the presence of prey, the hollow thread is ejected, being turned inside out in the process.

The largest type of nematocysts are the *penetrants* which have threads barbed at the base. These penetrate the prey and a paralysing fluid is injected. The *volvants* eject smooth threads which entangle hairs and bristles of the victim. The two kinds of *glutinants*, one larger than the other, have sticky threads which do not seem to be used in capturing prey, but as a means of attaching the hydra to a support when it is moving about. The poisonous nematocysts can be used only once, but new ones develop to take the place of those that have been expended. (Plate **16, 3**)

Only animal food is taken by a hydra, and this consists mainly of small crustaceans such as *Daphnia*, or aquatic worms, but small fish can be taken. Hydras are sometimes troublesome in aquaria containing fish-fry. Once inside the body of the creature, the food is digested by cells of the inner layer which line the gut-cavity. These digestive cells are of two kinds. There are some that can put out blunt arms like the pseudopodia of an *Amoeba* and engulf small particles. When these cells are not engulfing food, the arms are withdrawn, and in their place hair-like cilia are put out which, by their waving action, keep the fluid in the body-cavity in motion.

Other cells secrete a digestive liquid which acts on the larger pieces of food and reduces them to a fluid form capable of being absorbed. Waste products and indigestible matter are carried to the mouth opening in the current caused by the lashing of the cilia, and are evacuated from the body.

The minute algae which give hydras their colour are to be found in the digestive cells. They benefit from the nitrogenous and other waste products given out by their host in its metabolism, and utilize its carbon dioxide, providing in return a certain amount of oxygen by their photosynthesis. No special respiratory organs exist in hydra and the intake of oxygen and the release of carbon dioxide take place through the cell-wall all over the body.

The appetite of a hydra is almost insatiable. When food is plentiful a truly amazing amount is caught and digested, and the body becomes very much distended. At such times one or more buds appear on the side of the body, and grow rapidly, soon developing tentacles of their own and resembling the mature animal in everything except size. The bodies of the new hydras are hollow, and communicate with the gut-cavity of the parent but, nevertheless, they catch food on their own account, sometimes even competing with the adult for a particular morsel. In time the point of attachment of the bud constricts, and soon the young hydras break free to lead independent lives. (Plate **16, 4**)

Hydras also reproduce sexually, usually in the autumn. Male and female cells develop sometimes on the same individual and sometimes on separate ones. The male sperm is in small bulges near the mouth end and the single eggs are in the larger swellings nearer the base. (Plate **16, 2**) The sperms escape into the water and, by lashing action of their flagella, swim to the swelling containing the egg-cell; one of them fertilizes it. The fertilized egg becomes surrounded by a tough covering, and it is then released from the body. The eggs of the green and common hydra drop to the bottom of the water, but those of the brown hydra become attached to water-plants. In spring, larvae emerge from the eggs, which quickly develop into mature hydras.

Hydras can also multiply in another way, although it is doubtful if the method is used more than very occasionally in a natural state. The Abbé Trembley, who was the first to study the creature—the Freshwater Polyp as he called it—about the year 1750, found that if he cut a hydra into a number of pieces, each grew into a new hydra. It was not possible to continue the process indefinitely and, as we know now, the pieces had to contain both ectoderm and endoderm before they would give rise to new individuals. Even so he showed that the animal had the ability to regenerate itself from a separated part of its body. On account of this property, Trembley called the creature hydra, after the monster in Greek mythology with a hundred heads, which grew two new ones when one was cut off. The hydra was finally slain by Hercules, who hit on the idea of burning each fresh cut as he made it with a hot iron, and then poisoning his arrows in the monster's bile.

Trembley also claimed to have turned a hydra inside out, in which condition he maintained it continued to live quite happily. Sceptical present-day biologists accepted this story with a grain of salt, until an American biologist, with the advantages of modern techniques at his disposal, repeated the experiment in 1933 and confirmed Trembley's findings. This biologist discovered that the cells of the inner and outer layer migrated so that they once again regained their normal positions relative to each other.

Although hydras usually remain firmly attached by a sticky secretion to their supports for long periods, swaying this way and that, and waving their tentacles around freely in search of food, they can and do move from place to place when necessary. Sometimes short movements are made by gliding along the support. At other times they indulge in a form of swimming, first releasing their hold, and then writhing about in the water before settling in a new position. But probably the most usual mode of progression is by gripping a surface with the tentacles and mouth, detaching the base, and then moving the body over to a new point of attachment, repeating the process time after time.

Sometimes when a hydra is examined under a medium power of the microscope, small creatures will be seen moving about on its body. These are ciliated Protozoa, related to *Vorticella*. One shaped rather like a vase or barrel is *Trichodina polyporum*; a flatter type, in general shape like *Paramecium*, is called *Kerona pediculus*. Both these organisms live as external parasites on other creatures without apparently causing their hosts any inconvenience. It is believed that in addition to the scavenging role of *Kerona* in removing debris, bacteria, desmids and diatoms from the body of hydras, it is dependent on them for a chemical secretion. The crustacean, *Anchistropus emarginatus* (p. 148), feeds on the ectoderm of hydras but rarely seems to kill them.

It is interesting that all these external parasites do not trigger off the hydra's defensive nematocysts. The flatworm, *Microstomum lineare* (p. 95), is not always so fortunate, but if it can attack a hydra at its base it not only avoids any ill

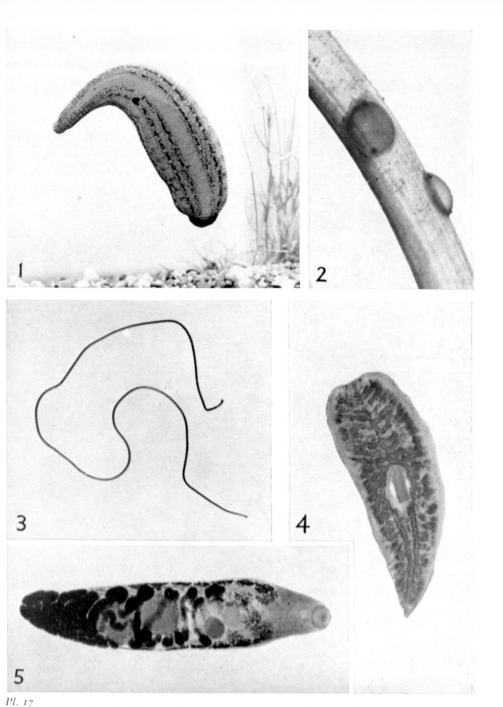

Pl. 17

1 Medicinal leech, *Hirudo medicinalis* (× 1). 2 Cocoons of *Erpobdella octoculata* on the stem of a submerged plant (× 3). 3 Hairworm, Gordiidae (× ½). 4 Planarian flatworm, *Dendrocoelum lacteum* (× 4). 5 *Haplometra cylindracea*, a fluke found in the adult stage, shown here, in the lung of a frog (× 8)

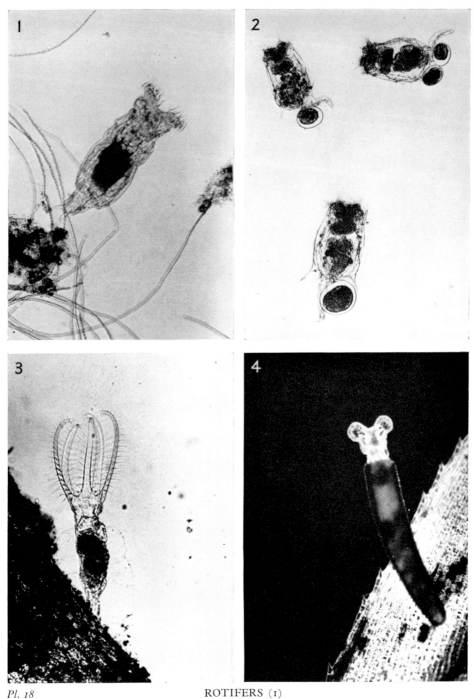

Pl. 18   ROTIFERS (1)

1 *Philodina roseola* (× 75). 2 *Brachionus* sp. with eggs (× 75). 3 *Stephanoceros fimbriatus* (× 50). 4 *Floscularia ringens* (× 40)

Pl. 19 MOSS ANIMALS (1)

1 Colony of *Plumatella* on a tree rootlet (× 3). 2 Colony of *Lophopus crystallinus* (× 12)

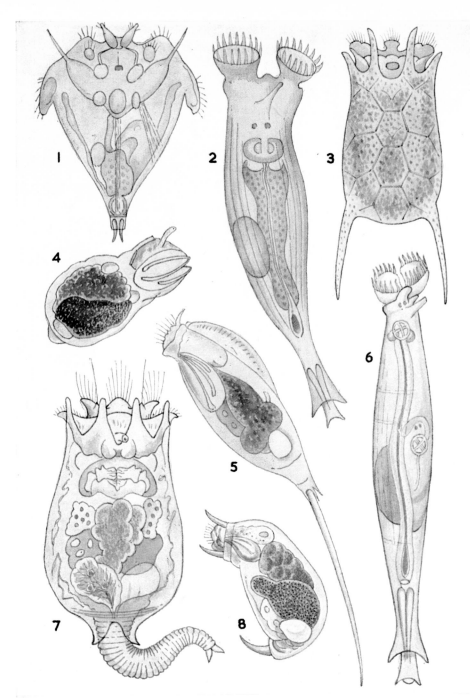

Pl. 20

ROTIFERS (2)

1 *Synchaeta pectinata* (× 175). 2 *Philodina roseola* (× 130). 3 *Keratella quadrata*
(× 275). 4 *Ascomorphella volvocicola* (× 220). 5 *Trichocerca cristata* (× 275). 6 *Rotaria
rotatoria* (× 175). 7 *Brachionus calcyiflorus* (× 175). 8 *Trichocerca porcellus* (× 130).

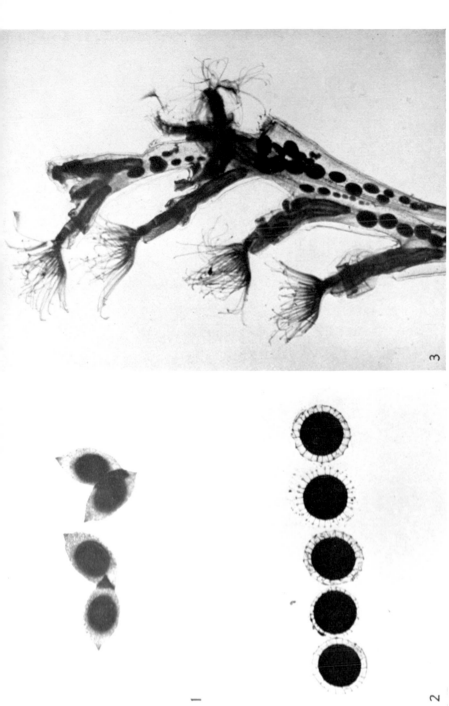

Pl. 21

MOSS ANIMALS (2)

1 Statoblasts of *Lophopus* (× 10). 2 Statoblasts of *Cristatella* (× 10). 3 *Plumatella* with statoblasts (× 5)

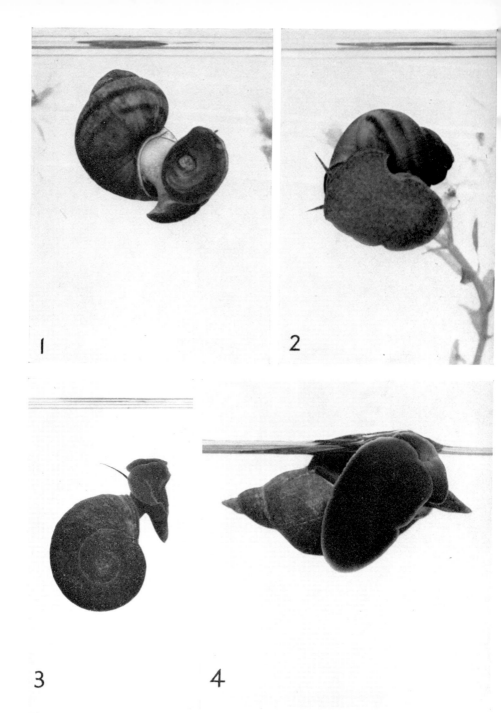

Pl. 22

1 Freshwater winkle, *Viviparus viviparus*, male, showing operculum (× 2). 2 *Viviparus viviparus*, female (× 2). 3 Great ram's-horn snail, *Planorbarius corneus* (× 1). 4 Great pond snail, *Lymnaea stagnalis* taking air at the water surface (× 2)

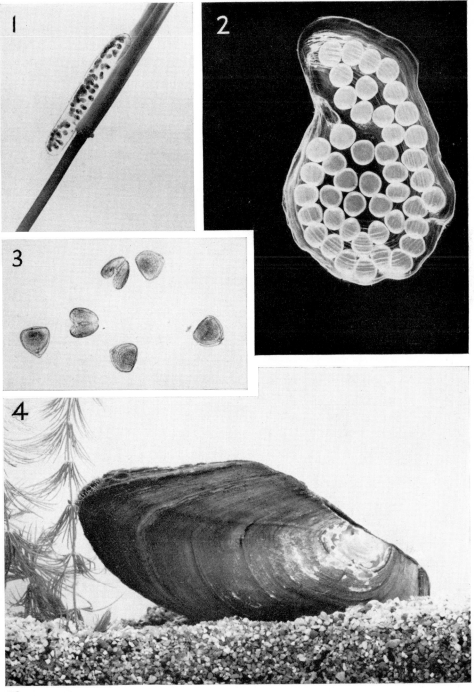

Pl. 23

1 Eggs of great pond snail, *Lymnaea stagnalis*, on a stem of *Potamogeton* ($\times$ 1½). 2 Eggs of great ram's-horn, *Planorbarius corneus* ($\times$ 4). 3 Glochidium larvae of swan mussel, *Anodonta cygnea* ($\times$ 20). 4 Swan mussel, *Anodonta cygnea*, showing the exhalant and in-halant siphon openings on the upper left side of the shell ($\times$ 1)

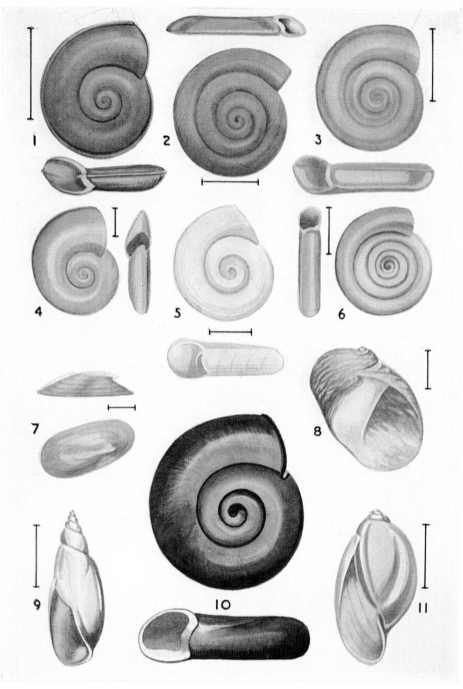

Pl. 24

WATER SNAILS

1 Keeled ram's-horn, *Planorbis carinatus*. 2 Whirled ram's-horn, *P. vortex*. 3 The ram's-horn, *P. planorbis*. 4 Flat ram's-horn, *P. complanatus*. 5 White ram's-horn, *P. albus*. 6 Button ram's-horn, *P. spirorbis*. 7 Lake limpet, *Ancylus lacustris*. 8 The nerite, *Theodoxus fluviatilis*. 9 Moss bladder snail, *Aplecta hypnorum*. 10 Great ram's-horn, *Planorbarius corneus*. 11 Bladder snail, *Physa fontinalis*. (Scale lines twice natural size)

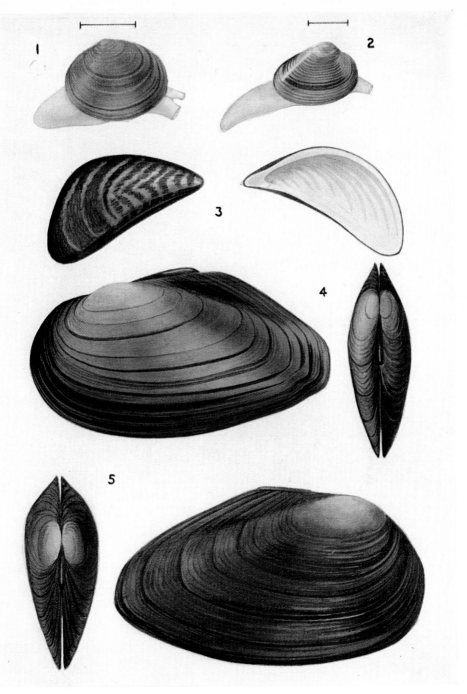

Pl. 25 1 Horny orb-shell, *Sphaerium corneum*. 2 River pea-shell, *Pisidium amnicum*. 3 Zebra mussel, *Dreissena polymorpha*. 4 Swan mussel, *Anodonta cygnea*. 5 Duck mussel, *Anodonta anatina*. (Slightly reduced, except 1 and 2, where the scale lines are twice natural size)

Pl. 26   CRUSTACEANS (1)

1 *Leptodora kindti* (× 10), found in the plankton of lakes and large reservoirs. 2 *Daphnia magna* with ephippium (× 10). 3 *Cyclops*, female with egg-sacs

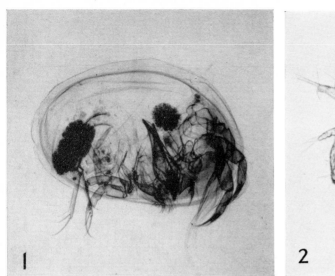

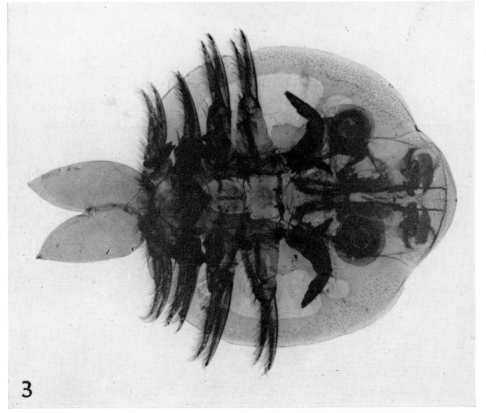

Pl. 27
CRUSTACEANS (2)

1 *Cypria*, treated to show the internal structure (× 25). 2 *Cyclops*, nauplius larva (× 100). 3 Carp louse, *Argulus coregoni*, ventral aspect (× 15)

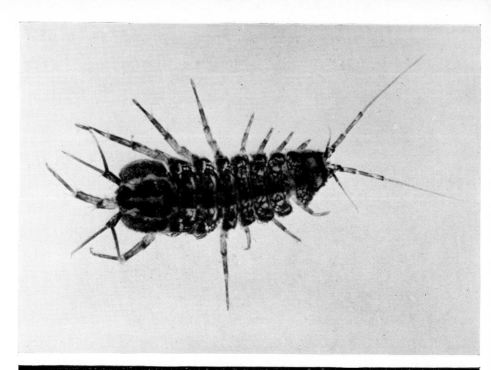

Pl. 28 CRUSTACEANS (3)

Above: Water louse, *Asellus aquaticus* (× 4). *Below:* Freshwater shrimp, *Gammarus pulex* (× 4)

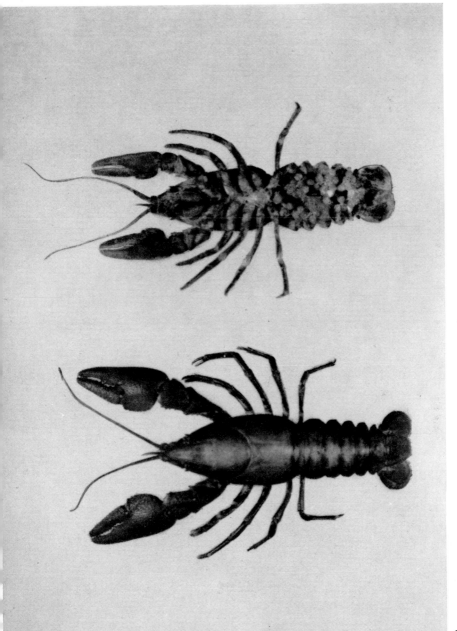

Freshwater crayfish, *Astacus pallipes*: *right*, female underside, to show the eggs attached to the swimmerets. (Slightly reduced)

Pl. 29

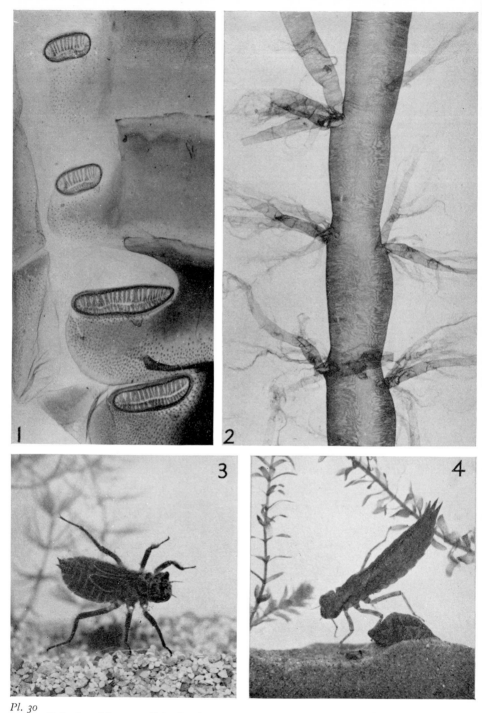

Pl. 30

1 Spiracles of the great diving beetle, *Dytiscus marginalis* (× 20). 2 Part of main trachea or air tube of *Dytiscus marginalis* (× 20). 3 Plump, short-bodied libellulid type of dragonfly nymph (× 1½). 4 Long-bodied anisopterid type of dragonfly nymph (× 1)

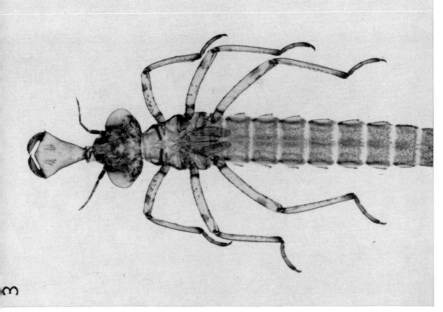

Pl. 31

1 Damselfly just emerged from the nymphal case, which can be seen on the right ($\times$ 1½). 2 Nymph of damselfly ($\times$ 3½). 3 Forepart of body of damselfly nymph, showing the 'mask' outstretched, and also the wing-buds ($\times$ 9)

DAMSELFLIES

1 Banded agrion, *Agrion splendens*. 2 Common blue damselfly, *Enallagma cyathigerum*. 3 Small red damselfly, *Ceriagron tenellum* (= *Pyrrhosoma tenellum*). 4 Demoiselle agrion, *Agrion virgo*. 5 Green Lestes, *Lestes sponsa*. 6 Common Ischnura, *Ischnura elegans*. 7 Large red damselfly, *Pyrrhosoma*

Pl. 32

effects of the stinging cells, but it absorbs them into its own tissues and uses them for its own defence.

A hydra is altogether a fascinating creature to study, and it illustrates in a simple way the intricacy of the higher many-celled animals. In the specialization of its cells into skin-cells, stinging cells, primitive muscles, nerve-cells, cells secreting digestive fluids, eggs and sperms, it represents a distinct advance in development over the primitive creatures discussed so far, yet the amoeba-like digestive cells remind one that it has affinities to the protozoan type of organism from which it has evolved. It will come as no surprise to the reader that the hydra is now one of the most important research animals in use today and is playing an important part in many fields of study, including cell biology, symbiosis, behaviour, growth and regeneration, and parasitology.

In many marine hydroids, two stages are met with: the fixed hydra-like creature, and a free-swimming transparent *medusa*, shaped rather like a bell. These medusae arise as buds on the side of the fixed form, assume their character-istic jelly-like appearance, and then become detached and swim away. They are reproductive stages, sperms being formed in some and egg-cells in others. After fertilization, the egg-cells develop into small free-swimming organisms, which escape from the medusa and finally settle on a solid object and grow into a hydroid form once again. Much to the surprise of the scientific world, fresh-water medusae were found in 1880 in the tank containing the giant water-lily *Victoria regina*, at what were then the gardens of the Royal Botanical Society in Regent's Park, London. Owing to some confusion, the animals were given two scientific names on the same day by leading scientists: *Limnocodium victoria* and *Craspedacusta sowerbii*. The first was used in accounts of the creature until 1910, when the International Commission on Zoological Nomenclature ruled that *Craspedacusta sowerbii* had priority; with the specific name changed to *sowerbyi*, this has been its correct name ever since.

Four years after finding the medusae, the polyp stage was discovered on the roots of the plant *Pontederia*, when a warm water tank was being emptied in the same gardens. This polyp was given the name *Microhydra ryderi*. It may seem strange that different names should be given to what are merely two stages in the life-history of one animal, but at first the connection between the two was not clear. The same situation has arisen in other animals, including gall insects, lampreys and eels.

Microhydra resembles a small hydra and is only 0·5–2·0 mm long, with no tentacles, although there are nematocysts surrounding the mouth. It may grow singly, but colonies of up to seven polyps have been seen. In winter, the hydroid contracts into a solid mass of cells, surrounded by a horny coating. In this state it can survive during adverse conditions and then redevelop into a hydroid when more suitable weather returns.

The medusa stage is released from buds on the hydroid and at first measures

G

only about 0·5 mm, both across the umbrella and in height. It has eight tentacles, armed with stinging cells, around the margin of the bell, but later sixteen develop. Eventually, when fully grown, the animal may measure as much as 20 mm across and bear 480 tentacles, under optimum conditions. The medusae swim actively by making vigorous contractions of the umbrella, and they tend to congregate in well lit situations. They feed on a variety of small freshwater creatures including crustaceans, rotifers and Protozoa. Cases are on record where fish have been attacked; one goldfish lost part of its tail-fin as a result of contact with the stinging tentacles of a medusa.

In Britain, a number of both medusae and polyps have been recorded in recent years. These have usually been found in heated aquaria, or in outdoor waters such as canals heated by industrial processes. However, occurrences have

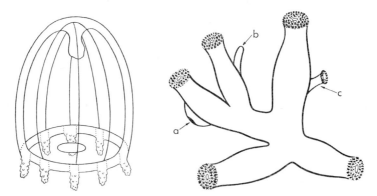

Fig. 35 Left: young medusa of *Craspedacusta sowerbyi* (× 90). *Right:* small colony of *Microhydra ryderi*: a. developing medusa, b. and c. stages in development of polyps (× 15)

been reported in the Exeter Ship Canal, a colliery reservoir in Monmouthshire and Witcombe Reservoir in Gloucestershire. These creatures were found in the summer months when the water temperature was about 20° C. It is clear that the animals originate in warmer countries, and it is known that they have a wide distribution throughout the world. It seems certain that they are introduced to this country on imported water-plants in the hydroid stage, and it has been observed that they are usually of one sex only when found, and thus can give rise only to medusae of one sex by asexual budding. They probably occur more frequently than the few records indicate, and are overlooked.

The medusae seen in Britain are very small and usually do not develop further than the stage with eight tentacles. Most of the recorded occurrences have been during the months of June to September. Readers who have tropical aquaria may well see them, if they examine their tanks at these times in the best lit part, as small free-floating umbrella-like objects from 0·5 mm to 3 mm across. They are most clearly seen against a dark background.

Cordylophora

If we imagine a hydra in which the buds grow out on stalks and remain attached to the parent, a good idea will be obtained of how the colonial forms of Hydrozoa are formed. There are many marine genera which adopt this colonial habit, often secreting a tough, protective covering over their common stem, extending, in some species, right to the tentacles of the individual polyps, and in others stopping at the base of the polyps.

A brackish-water species of this colonial type, *Cordylophora lacustris*, sometimes occurs in fresh water. It forms masses on submerged objects such as the uprights of landing-stages and bridges. (Plate **16**, 1)

8: FLATWORMS AND ROUNDWORMS

The word 'worm' is generally used somewhat loosely to mean almost any creature which is long and narrow in shape, and so, not only the common earthworm, but leeches, thread-worms, fluke-worms, planarian worms and many more are all comprised in this convenient, if unscientific, category. It is unscientific because there is no close relationship between the three main phyla: flatworms, roundworms and true or segmented worms.

It may be appropriate, however, to give here a few remarks on 'worms' generally, for they do show certain important advances in development over the simpler many-celled animals which were dealt with in the last two chapters. In the first place, their bodies have a third layer of cells between the outer ectoderm and the inner endoderm. It will be remembered that the sponges and hydra had merely a jelly-like non-cellular substance between their two layers of cells. In the worms, and in all the phyla of animals of higher development, culminating in man, this layer of jelly is replaced by a middle layer of cells called the *mesoderm* (Gr. *mesos*, middle; *derma*, skin). It is from the mesoderm that *tissues*, such as muscles, and associations of tissues forming *organs*, are derived, and these make possible an increasing complexity and efficiency in animal activities. On account of their three layers, all animals from worms onwards are called *triploblastic* (Gr. *triplōos*, L. *triplus*, triple; *blastos*, a bud, shoot); sponges, and hydra and its relatives are called *diploblastic* (Gr. *diplōos*, double).

The worms show a further stage in development by having what is known as *bilateral symmetry*, which means that their body structure is arranged on two sides of an imaginary plane running from the front end of the animal to the rear end. The hydras and their relatives, on the other hand, have *radial symmetry*; they have no left or right sides, and are the same all the way round. A downward slice taken through their centre in any direction would produce two parts of the same shape. Bilateral symmetry lends itself to a form of body capable of easy movement, and gives rise to active and successful animals capable of searching for their food or mates, and of escaping from their enemies. It makes a definite step forward from the fixed or drifting type of many-celled animals which have to wait for food to come within reach.

In some of the higher groups of 'worms' there is the beginning of a centralized nervous system, with a 'brain' controlling the activities of the creatures; a true body-cavity or space between the food-canal and body-wall also appears for the first time.

The human race, as a whole, seems to have a repugnance to 'worms', and perhaps to this fact can be attributed the lack of attention which the average pond-hunter gives to the various groups which are found in fresh water. Yet they will repay more careful study, not only as examples of the first active many-celled animals, but also on account of the almost unbelievable life-histories through which some of them pass, and their importance in the lives of other creatures.

Phylum PLATYHELMINTHES

(Gr. *platys*, flat; *helminth-*, a worm)

Flatworms

The flatworms are the simplest as regards development of the worms, with an unsegmented body which, as their name implies, is flattened, so much so that they have no body-cavity, and their organs are embedded in the compact middle layer of tissue of the mesoderm. Nearly all of them are hermaphrodite, that is having both male and female organs in the same animal. They are, therefore, each capable of producing both eggs and sperms, but the eggs of one individual must be fertilized by the sperms of another.

There are three classes of flatworms: Turbellaria, which includes the planarians, Trematoda (flukes) and Cestoda (tapeworms).

Class *TURBELLARIA*

The turbellarians (L. *turbella*, a small disturbance) were so named because of the currents caused in the water surrounding them by the vibrating cilia with which their body, particularly the under-surface, is covered.

Although a medium power of the microscope is needed to see the actual cilia, if a planarian is examined under a × 10 objective, the effect of the turbulence it produces can be seen by the way minute particles in the water are wafted about. The currents created are to aid respiration, and the movement of the cilia also enables the creature to glide smoothly over any surface, even the underside of the surface film of the water. Slime secreted by glands just under the surface of the body lubricates the surface over which it is moving. At times, movement is speeded up by muscular spasms of the body, but flatworms are incapable of swimming freely, except when very small. Most turbellarians have large numbers of small, dark, rod-like crystalline bodies in the cells of the skin, called *rhabdites*. These, when discharged into the water, disintegrate into a thick, sticky mass, and are thought to be used as means of defence and attack.

The planarians are the largest of the freshwater turbellarians, and are well known to all pond-hunters. The species vary in size from 8 mm to 35 mm, and in colour from white through grey and brown to black. Ten species are found in Britain, and they occur in both still and running water. They live a free existence, and are not parasitic on other creatures. Their bodies are soft, elongated and greatly flattened, and may have short tentacles at the head end; eye-spots, varying in number according to the species, can be detected in a close examination. Planarians are largely nocturnal creatures and tend to shun the light, being found usually during the day-time under stones or on the underside of leaves of plants, looking like small lumps of jelly. At times a piece of water-plant taken out of the water will be found swarming with the creatures, or a stone upturned may reveal two or three dozen. The best way of removing them from their support is with a paint-brush. A piece of fresh meat or liver left in the water for a short time will usually attract specimens for study.

Planarians belong to the order of turbellarians known as the Tricladida (Gr. *treis*, three; *klados*, a branch), characterized by the way in which the alimentary

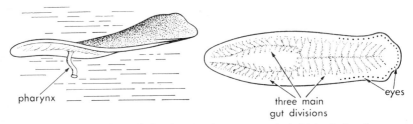

pharynx

three main
gut divisions

eyes

Fig. 36 Planarians: *Left:* with the pharynx extended. *Right: Polycelis,* showing the branched alimentary canal and the eyes

canal is divided into three main branches. The mouth is not situated at the head end, but about the middle of the underside of the body, and through its opening can be protruded at will a tube-like pharynx. Planarians feed on almost any small creatures, or even on dead animal matter, and are able to detect the presence of food from some distance away by means of two olfactory organs in grooves near the head. When prey is located, the pharynx is protruded and, by a sucking action, the solid food is torn into small pieces and swallowed, together with the victims' juices. Planarians are able to exist for long periods without food, but they become smaller and smaller as time goes on, through the digestion of their own tissues.

The mouth serves not only to take in food, but for the excretion of waste products as well. Planarians also have, running the whole length of their bodies, a network of fine tubes opening to the surface here and there by pores. From the tubes branch off little swellings each containing a tuft of cilia, which by their movement cause the circulation of fluid along the tube network to the pores. From the resemblance of the movement of the cilia to the flickering of a flame, these cells have become known as 'flame-cells'. It is believed that the currents they create are primarily for the regulation of the water content of the tissues.

It has already been mentioned that each flatworm has both male and female organs within it, but the union of two individuals is necessary before the eggs of either can be fertilized. The eggs of planarians are usually laid, a number together, in a cocoon, often attached to submerged stones or plants. The cocoons are either spherical or oval depending on the species; those of *Dugesia* have a short, thin stalk. Cocoons change colour from yellow or orange to dark brown within a few hours of being deposited. The eggs are unusual in that the yolk does not occur in the egg itself, but in special yolk-cells which accompany them. The time of egg-laying varies among the different species of planarians, some choosing the spring or early summer, whereas others apparently breed all the year round. There is no larval stage, and from the eggs hatch little planarians resembling their parents, except of course in size.

Reproduction by eggs is not the only method of multiplication adopted by planarians; they have remarkable powers of regenerating themselves, and should one be severed accidentally, or for the purpose of experiment, a new animal will grow from each piece. Some species, in fact, habitually multiply by having a tug-of-war with themselves, the two pieces which separate as a result becoming complete worms.

The flatworms are the simplest animals to have a central nervous system such as is possessed by all higher animals. A very primitive brain is present in the head region, and from it strands of nerve-cell ramify throughout the length of the body. The creatures are, therefore, very responsive to outside influences, and many interesting experiments can be carried out to test their reactions to such stimuli as light, variations of temperature, food and water currents. If, for instance, a planarian is placed in a dish of water, it will turn away from the light and move towards the shadiest part of the dish. It has been shown that this happens even in planarians whose eyes have been removed, which indicates that light-sensitive cells are present in other parts of the body. Then, again, some planarians move towards a current in the water, in the same way as in their native brook they would tend to move upstream. Chemical substances in the water affect them in various ways. If a small piece of raw meat is placed in the dish containing them, they will rapidly move in its direction guided, not by sight, but by the impact of the meat juices on sensory cells of the body. Other substances will repel them.

The identification of planarians is not easy, based as it is largely on the structure of the sex organs and the arrangement of the eyes, both features which are not easily seen at the waterside. The different species assume characteristic shapes when moving freely, and the colours, although varying even between individuals of the same species, may serve as a guide.

The largest British species is *Bdellocephala punctata*, of which an extended

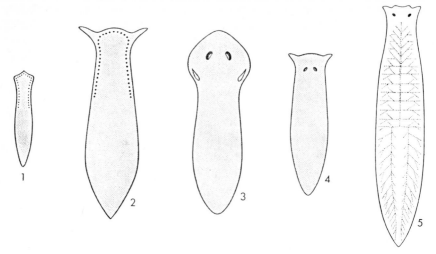

Fig. 37 Planarians: 1 *Polycelis nigra,* 8–12 mm. 2 *Polycelis felina,* up to 25 mm, variable. 3 *Dugesia lugubris,* 11–17 mm. 4 *Crenobia alpina,* about 12 mm, variable. 5 *Dendrocoelum lacteum,* 10–14 mm

adult may be 30 mm in length and of a brownish colour. It usually lives on the underside of stones in lakes. *Dendrocoelum lacteum,* a very common species, is, as its specific name implies, milk-white and may be as much as 25 mm long. (Plate **17,** 4) It is especially abundant in productive lakes where the freshwater louse, *Asellus,* is present, for this crustacean provides an important part of its food. *Crenobia* (= *Planaria*) *alpina,* usually greyish, varies greatly in colour from white to black, and is much smaller than the two preceding species, full-grown specimens being only about 12 mm long. This worm occurs only in cool waters such as mountain brooks or springs, moving farther upstream, where it is cooler, in summer, and returning to lower reaches in winter.

Dugesia lugubris, black or dark brown, has a white area surrounding each of its two eyes which gives it the doleful expression denoted by its specific name. Recent surveys have shown that the very similar species *D. polychroa* is more widespread in Britain than *D. lugubris,* and most of the earlier records of the latter species probably refer to *D. polychroa,* which is slightly smaller, but with a more rounded head, eyes placed further back and a more pronounced neck. It is common on the underside of stones and plants in productive lakes and feeds largely on water-snails.

Polycelis nigra, another very common species in lowland streams and lakes, is usually black, but sometimes brown, and has a large number of eyes arranged round the front margins of its body. Specimens of up to about 12 mm are found. The egg-cocoons of this species, yellow at first, but changing to dark brown or reddish as they get older, are very familiar objects on submerged stones or plants in spring. *Polycelis tenuis* is common in ponds, lakes and slow streams, and can be distinguished only from *P. nigra* by examination of the penis, which is blunt in the latter species and elongated in *P. tenuis. Polycelis felina* (= *cornuta*), a large worm nearly 25 mm long, brownish in colour, is usually found under stones in small streams. This species and *Crenobia alpina* are the only two which have tentacles. Since *Crenobia* has few eyes and *Polycelis* many eyes, these characteristics are useful identification features.

Planarians are occasionally troublesome but unsuspected pests in aquaria, including those used for tropical fish. Introduced, perhaps, on new plants which have not been sterilized, they remain hidden during the day under the stones, but at night sally forth and play havoc among the baby fish or ova which may be there.

Rhabdocoelida (Gr. = rod-like cavity) is another group of turbellarians with freshwater representatives, but they are mostly very small. In general shape they resemble planarians, but their alimentary canal is straight and not branched. They feed mainly on minute animal life, but also eat algae.

Dalyella viridis, about 5 mm long, is common in spring in small pools and ditches. The specimens that are green owing to the presence of symbiotic algae are easy to see, but many are colourless. *Mesostoma ehrenbergi* and *M. quadrangulare* are up to 15 mm long and therefore easiest to study, especially as they are almost transparent, although this makes them hard to see when collecting. They both occur in ponds.

Microstomum lineare, about 2 mm long, is usually found in chains of several individuals divided from each other by a slight 'waist'. This is a result of budding. The worm is interesting because it feeds on hydras and manages to absorb undischarged nematocysts from its prey, which are then conveyed to its own skin as a means of defence (p. 86).

One rhabdocoele, *Castradella granea,* is parasitic on the freshwater louse, *Asellus,* living in its brood-pouch.

Other species are found on algae or among the decaying plant and animal matter on the mud surface in still water.

Class *TREMATODA*

The Trematoda (Gr. = perforated, in reference to the cavity of the suckers which superficially resembles a perforation of the body), or flukes, are somewhat similar in general shape, in their adult stages, to planarians, but are readily distinguished by the presence of one or more suckers on the underside of their bodies, and by the absence of a covering layer of cilia. They are parasitic on other animals during the greater part of their lives, and by means of the suckers are enabled to cling to their hosts, sucking the juices. In the middle of the front sucker is the mouth, which leads by way of a pharynx into a much-branched alimentary system. Eyes or other sense-organs are rarely present in the adult. Male and female organs, frequently of great complexity, are found in each individual, and flukes are unlike the planarians in that self-fertilization often takes place. Some produce eggs in considerable numbers, but others bring forth their young alive.

There are three orders of trematodes: Monogenea, with a comparatively simple life-history and parasitizing only one host, usually externally; Digenea, with a complicated life-history, involving two or more hosts; and Aspidogastrea, some of which have one host and some two.

MONOGENEA

Members of the first of these orders are those most likely to be encountered in fresh water, for belonging to it are the creatures known as gill-worms, or gill-flukes, which are often attached externally to fish, particularly to the gills. Sometimes when a fish is examined, it will be found literally swarming with the

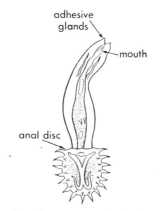

Fig. 38   Gyrodactylus (× 60)

creatures, as many as two or three hundred having been seen on a single three-spined stickleback. These are of the genus *Gyrodactylus*, the adults being about 1 mm long, with the greatest width in the middle part of the body. The young, which may be seen inside the adult, are born alive, and attach themselves almost immediately to the same host as their parent. *Dactylogyrus*, although not so far recorded in a wild state in this country, has been found in aquarium fish (as has also *Gyrodactylus*), and has presumably been brought over from the Continent, where it occurs on carp and bream. The young emerge as larvae from eggs laid singly, and swim round with the aid of their cilia until they find the appropriate host. *Diplozoon* infests various freshwater fishes, including the carp family and the stickleback. The eggs are laid on the gills of the fish, and on hatching the larvae swim round in the same way as *Dactylogyrus* until they find their host. If they fail to do so in five or six hours, they die.

Polystoma occurs in the bladder of frogs. Its eggs, numbering as many as 1,000, are laid in spring, and reach the outside water through the urinary aperture of the frog. The larvae attach themselves to tadpoles and make their way into the creatures' branchial chambers. When the tadpoles turn to frogs the larval *Polystoma* work their way into the intestine, and then into the bladder, where they remain three years before becoming mature. Should a larval *Polystoma* attach itself to a newly hatched tadpole which has external gills, the richer blood supply there causes it to develop into an adult in as short a time as five weeks. In this case there is no penetration to the interior of the frog, and the fluke, having discharged its eggs, dies.

DIGENEA

Flukes of the second order are all internal parasites of other animals, living in the liver or other organs of their hosts. The best known is the liver fluke, *Fasciola hepatica*, which causes fascioliasis or 'liver-rot' in sheep and cattle, as well as a wide variety of other hosts including pigs, rabbits, hares, coypu. The adult fluke is, of course, not aquatic, but the justification for discussing the animal in this book is that its larval stages are passed in the body of a common water-snail.

The bile-ducts of sheep in an advanced stage of infestation are often a writhing mass of adult flukes. Their eggs pass with the bile into the intestines of the sheep, and eventually outside the animal altogether. Here, among the damp grass in a marshy area, they hatch, and from them emerge a ciliated larval form called a *miracidium*, which either crawls about among the damp herbage or finds its way into the water. In either case, it moves about by means of cilia until it encounters a snail, and the usual species parasitized is *Lymnaea truncatula* (see p. 127). If a suitable species of snail is not found within a few hours, the miracidium dies.

Having reached its objective, a miracidium works its way into the snail's breathing aperture, and there settles down, losing its cilia and growing into a hollow sac called a *sporocyst*. Inside this a number of a second kind of larva called *redia* develop, and make their way into some other part of the snail, usually the liver. What happens then depends on the particular time of year. In winter, the rediae each give rise to more rediae, but in summer they give rise to a third kind of larvae with tails, which are called *cercariae*. These leave the body of the snail, and after swimming round for a time attach themselves to vegetation at the edge of the water, lose their tails and surround themselves with a protective covering. For many, this must be the end of their existence, but some, on blades of grass eaten by sheep, are swallowed and carried into the sheep's intestines, from which they reach the liver and develop into adult flukes.

The losses to agriculture due to fascioliasis in sheep and cattle are an increasing worldwide problem, and it is the commonest disease in animals at British slaughter-houses. The infected livers are condemned, although the parasites are at a stage when they could not cause fascioliasis to human beings. A much more likely source of infestation to man is when the cercariae have left the snail and encysted on wild watercress, *Rorippa nasturtium-aquaticum*, growing in ditches or ponds. If this is then eaten, the parasite will continue its life-cycle in human

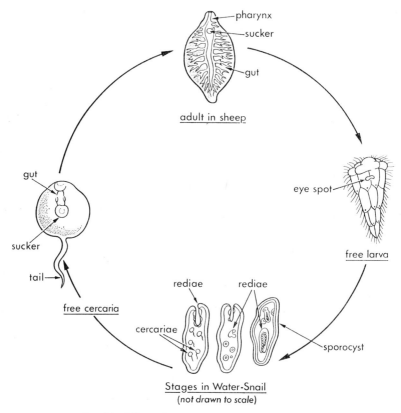

Fig. 39   Life-cycle of liver fluke, *Fasciola hepatica*

beings, the resulting disease being difficult to diagnose. In an outbreak recently reported by the Medical Officer of Health for West Gloucestershire, seven cases in five hospitals were at first wrongly diagnosed. The importance of eating only watercress grown in properly supervised beds, and bought in bunches bearing a producer's name, cannot be stressed enough. Washing infested watercress even in salty water will not remove the danger.

Both flukicides, which kill adult flukes within the hosts' livers, and molluscides, which kill the snail intermediate hosts, exist and are used, but are of somewhat limited value as few animals spend their whole life on one farm, and wild animals, such as rabbits, continue to provide host reservoirs. Drainage of land to eradicate the snail, or keeping animals off badly-drained areas seem the best chances of eradicating the liver fluke in domestic stock.

Other flukes which occur in fresh water and have even more complicated life-histories are *Bucephalus = Gasterostomum* and *Cotylurus*. The first is, in the adult stage, a parasite of perch and pike. The miracidium stage is passed in the freshwater mussels *Anodonta* and *Unio*. The cercariae swim round until they find one of the smaller fishes, such as a roach, when they encyst in its mouth. They can only complete their life-history and become adult if the roach is eaten by a perch or pike.

The miracidia of *Cotylurus* start their development in species of ram's-horn snails, *Planorbis*, and change to rediae and cercariae within them. Frequently leeches, particularly of the genus *Erpobdella*, feed on these snails, and in this way the cercariae gain access to the internal organs of the leeches, where they encyst. Before the trematodes can complete their life-history, however, it is necessary for the leeches to be devoured by a bird, probably some kind of water-fowl, in which the cercariae develop into adult flukes.

Haplometra cylindracea (Plate **17**, 5) is often found in the lungs of frogs. The cercaria stage takes place in a water-snail such as *Lymnaea stagnalis* and passes from there to the larva of a water-beetle, *Ilybius fuliginosus*, where it encysts. If the beetle, larval or adult, is eaten by a frog, the life-cycle is continued.

One cannot but marvel at the complexity of such life-histories as these, and particularly at the succession of coincidences which must take place before these creatures become fully adult and able to breed. Bearing in mind the apparently slender chances they have of survival throughout their development, we can hardly wonder at the fecundity which is such a characteristic feature of all of them.

ASPIDOGASTREA

The most familiar species of the third and last order is *Aspidogaster conchicola*, which lives inside freshwater mussels. It is about 3 mm in length, with a bladder-shaped body.

Class *CESTODA*

Tapeworms are exclusively parasitic, and in appearance are like a piece of wet white tape, often as much as several metres in length. The head (*scolex*) is provided with suckers and sometimes also with hooks, both of which serve to hold on to the tissues of the hosts, but they have no mouth, nor any trace of a digestive system of their own, merely absorbing the digested food of their hosts through their whole surface. Their bodies consist of a large number of sections which are continually being added to in the head region, so that the worm grows longer as it gets older. The sections each contain reproductive organs, and a truly enormous number of eggs are produced during the life of a single tapeworm, these eggs passing out of the body of the host in its excrement.

Adult tapeworms, with one exception, live in the alimentary canal of backboned animals, but their larval stages are passed in one or more intermediate hosts, some of which live in fresh water. *Schistocephalus gasterostei*, in the adult stage, lives in the digestive tract of birds. The eggs, on hatching near the water, develop into free-swimming, ciliated larvae which bore their way into small aquatic creatures such as *Cyclops*, where they develop and form a cyst. They do not develop any further unless the *Cyclops* is eaten by a fish, such as a stickleback, when they become active again, and eventually form a further cyst in the body-cavity of the fish, usually so large that it causes a bulge on one side of the body. To complete the life-history of the tapeworm, the stickleback must be eaten by a bird.

Diphyllobothrium latum, a related cestode, has a similar life-history and passes intermediate stages in *Cyclops* and freshwater fish, including pike. Human beings eating these infested fish become the final hosts and acquire the tapeworm, which may be up to 15 metres long.

Not all tapeworms need a host on land, and some complete their full development entirely in freshwater animals. Knowledge of the full life-histories of some species is incomplete, and much work on these interesting, if repulsive, creatures remains to be done.

Phylum ASCHELMINTHES

(Gr. *askos*, a bag; *helminth-*, a worm)

This phylum has been created to include four classes of animals which were formerly accorded the status of separate phyla:

NEMATODA   roundworms
NEMATOMORPHA   hairworms
ROTIFERA   rotifers
GASTROTRICHA   'hairy backs'

Roundworms: Class *NEMATODA*

(Gr. *nema*, a thread; *odes*, like)

We need do little more than mention the roundworms for, although both on land and in water these creatures are very abundant, some being of great importance as parasites of man and other animals, comparatively little is known about those found in fresh water.

Roundworms have long, thin, colourless, unsegmented bodies generally pointed at both ends and enveloped in a tough cuticle. The sexes are usually separate.

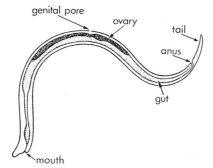

Fig. 40 Nematode, *Dorylaimus stagnalis* (× 100)

Many are parasitic for at least part of their lives, some in animals and others in plants, but quite a number are free-living, and in this latter category are included some common in fresh water. They are very small, the largest, *Dorylaimus stagnalis*, being about 5–8 mm in length. Others, such as *Rhabdolaimus aquaticus*, are only about 1 mm long. They are usually found in the bottom mud of still water, particularly among the roots of water-plants. Their characteristic vigorous, lashing actions and movement in S-shaped curves distinguishes them.

Hairworms: Class *NEMATOMORPHA*

Worms more likely to be noticed than the true roundworms are the hairworms (Gordiidea) which, except for a single marine species whose larvae are parasites in Crustacea, comprise the class Nematomorpha (Gr. = form of a thread). From their close resemblance to thick hairs, and their habit of appearing suddenly in quite small bodies of water such as drinking-troughs, and even in domestic water supplies, the superstition arose that they were horse hairs that had fallen into the water and come to life! Their other common name, gordian worms, refers to their occurrence in masses so tangled as to suggest a gordian knot—the intricate knot fastening the yoke and pole of a wagon dedicated in Greek mythology to Zeus by Gordius, father of Midas. It was said that whoever untied the knot would inherit Asia. Alexander the Great severed it with his sword, and thus arose the meaning that to cut the gordian knot was to take drastic or unorthodox action to resolve a problem or impasse. (Plate **17**, 3)

Little study of the biology of hairworms has been carried out in Britain, and

Fig. 41   Larva of hairworm (× 300)

our knowledge is derived mainly from investigations on the Continent and in America. Adult hairworms range in length from 10 cm to 70 cm, and their colours vary from grey or brownish to black. They are found in freshwater habitats or even damp soil, from about October to May. The sexes are separate, and the males can usually be distinguished by a forked rear end. The adults do not feed and are merely the reproductive stage in the worms' development. Eggs are laid in long gelatinous strings attached to water-plants. The next stage in the life-history is somewhat obscure, but it is believed that after a brief free-swimming period of about a day the larvae encyst on vegetation at the water's edge. These cysts remain viable for up to two months, and when the cysts or vegetation holding them are eaten by a variety of host insects the larvae, about 0·25 mm long, are released and bore through the gut-wall of the hosts into the body-cavity, where they feed on the surrounding tissue. Hosts include grass-hoppers and crickets, terrestrial and aquatic beetles, caddis-flies and dragon-flies. Cysts may, of course, be eaten accidentally by other creatures such as snails, worms and fish, in which case they do not develop but may re-encyst, and either die or continue their development when the abnormal host itself is eaten by a normal one. Metamorphosis to the adult hairworm within the host takes several months, presumably being dependent on favourable conditions for development. There is some evidence that terrestrial hosts tend to seek water when their parasites are about to emerge, and any newly-emerged worms soon die if they do not have access to water. If the worms emerge from their hosts in spring or summer, reproduction takes place soon after, but when emergence takes place in autumn they hibernate in damp surroundings until the following spring.

Four species of hairworm are widely distributed in Britain: *Gordius villoti*, *Parachordodes violaceus*, *P. wolterstorffi* and *P. pustulosus*, but their identification is a matter for specialists.

9: ROTIFERS

The Rotifera (L. *rota*, a wheel; *fero*, to bear) or wheel-animalcules as they were called by the older naturalists, are minute creatures, ranging in size from less than 0·1 mm to about 2 mm, but the majority average about 0·25 mm. They are essentially freshwater creatures, only a few being found in the sea.

Some species remain fixed to water-plants throughout their life; some attach themselves to crustaceans and to other creatures in the water, and others are actually parasitic. The majority, however, are free-swimming, although even some of these anchor themselves temporarily when feeding. A sweep of a fine-meshed net in almost any stretch of water can hardly fail to capture some of the free-swimming forms, but on account of their size a close scrutiny of the water which has collected in the tube of the net may be needed before they will be seen. Many of them are smaller even than some Protozoa, and the rotifers were, in fact, formerly included with the Protozoa and other apparently simple animals into a heterogeneous group, the Infusoria, since they all appeared in 'hay infusions'. As mentioned on page 100 they are now considered a class—the largest of the phylum Aschelminthes.

But rotifers are far removed from protozoans; their bodies are many-celled and have a complex internal structure with reproductive, nervous and excretory systems, as well as a form of body-cavity distinct from the alimentary canal. Although the bodies consist of many cells, this does not mean that one will find cells with distinct cell-walls, for the structure of an adult rotifer is made up of special groups of cells called *syncytia*, which are masses of protoplasm without a definite wall and each containing a number of nuclei.

Some of the worms and molluscs have an odd kind of larva called a *trochophore*, usually oval or pear-shaped, which has circlets of cilia on various parts of its body. These larvae bear a striking resemblance to the general structure of some of the rotifers, and this has led biologists to believe that rotifers are fairly closely related to both worms and molluscs, perhaps representing an intermediate group, though somewhat of an offshoot from the main evolutionary tree.

There are very many genera and species of rotifers, differing widely even in their most obvious characteristics. They are usually cylindrical, with a head region, a middle portion or trunk, and a narrower, or tapering, foot, bearing two pointed 'toes' and paired pedal glands that secrete a sticky material which helps to anchor the animal when it is stationary. The body is enclosed in a thin

H

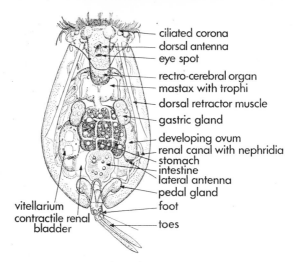

ciliated corona
dorsal antenna
eye spot
rectro-cerebral organ
mastax with trophi
dorsal retractor muscle
gastric gland
developing ovum
renal canal with nephridia
stomach
intestine
lateral antenna
pedal gland
vitellarium
foot
contractile renal
bladder
toes

Fig. 42   Structure of a rotifer, *Euchlanis dilatata*, dorsal
aspect (× 200)

elastic skin, sometimes hardened into a transparent shell, called a *lorica*, often
ornamented with ridges or other irregularities. The head region bears one or
sometimes two reddish eyes and the crown of cilia, which serves both as a means
of locomotion in the free-swimming forms and to create currents in the water,
which bring food and a constant supply of fresh oxygen.

In the first order of rotifers which was discovered—the *Bdelloida*—the cilia
are borne on expanded discs which, in their ceaseless waving, bear a striking
resemblance to the rotation of a wheel and gave rise to both the popular name
'wheel-animalcules' and also the scientific name Rotifera. The original description as given by Henry Baker in a letter to the President of the Royal Society in
1745 is worth repeating:

'I call it a *Water Animal*, because its appearance as a living creature is only
in that element. I give it also for Distinction sake the Name of *Wheeler*, *Wheel
Insect* or *Animal*; from its being furnished with a Pair of Instruments, which
in Figure and Motion, appear much to resemble Wheels. . . .

'A Couple of Circular Bodies, armed with small Teeth like those of the
Balance-wheel of a Watch, appear projecting forwards beyond the Head, and
extending sideways somewhat wider than the Diameter thereof. They have
very much the Similitude of Wheels, and seem to turn round with a considerable Degree of Velocity, by which means a pretty rapid Current of
Water is brought from a great Distance to the very Mouth of the Creature,
who is thereby supplied with many little Animalcules and various Particles of
Matter that the Waters are furnished with. . . .

'As the Animal is capable of thrusting these parts out, or drawing them in,
somewhat in the Way that Snails do their Horns, the Figure of them is

different in their several degrees of Extension and Contraction, or according to their Position to the Eye of the Observer, whereby they not only appear in all the various forms before represented, but seem at certain Times as if the circular Rim of the Wheel or Tunnel were of some Thickness, and had two Rows of Cogs or Teeth, one above and the other below that Rim.'

The species which Baker was describing is that now known as *Philodina roseola*, which will be mentioned below.

Another characteristic feature of the rotifers is the cavity known as the *mastax* near the front end of the alimentary canal. This is a form of gizzard with thick muscular walls, and is provided with a set of jaws or *trophi*, which are unique in animals. The form of the mastax varies in the different groups of rotifers and it is the principal feature used in their classification. The microscopic food, such as algae and smaller Protozoa brought to the mouth by the action of the cilia, are swept into the mastax, and there ground by the jaws into a pulp before passing on to the stomach for digestion. Waste matter passes out through an anus situated near the rear end of the body.

The passage of the food through the body can easily be studied in rotifers, for their bodies are sufficiently transparent to allow the internal organs to be seen. The process can best be watched in one of the fixed forms, particularly if it is feeding mainly on plant foods, the green colour of which shows up clearly.

Nearly all the rotifers that are found are females, and during most of the year the development of eggs proceeds without fertilization by a male, to produce further females. The eggs may hatch internally or be carried, until hatching, externally at the rear end of the body. Other species attach their eggs to plants or even other animals, and some merely drop them into the water to sink or float at the surface. Most rotifers produce two kinds of parthenogenetic (Gr. = virgin-born) eggs: large ones that hatch into females and smaller ones that will hatch into males. The latter are produced under adverse conditions such as cold weather, drought or overcrowding. The males mate with some of the females, and large eggs are laid which usually have a tough resistant shell, often with spines. These eggs tide the rotifers over the unfavourable time, and hatch later into females, which start the normal life-cycle again.

Male rotifers are very degenerate creatures, much smaller than the females, possessing no digestive organs and being little more than swimming sacs of sperms. In some species males have never been found and probably do not exist.

Some kinds of rotifers possess remarkable powers of withstanding drought or other unsuitable conditions. They retract into a compact form and surround themselves with a tough, protective skin. In this state they can exist for months, or even years, and may be dispersed widely by the wind. As soon as they reach suitable water they resume their former shape and activity. The sudden appearance of rotifers in new stretches of water and also in 'hay infusions' is no doubt due partly to this remarkable property which they share with other minute creatures, although other species are probably transported in the form of air-borne eggs.

The serious student of the rotifers soon comes face to face with the problem of nomenclature. In probably no other group of animals has there been so much confusion in giving scientific names as in the Rotifera. Many of the commoner species have had fifteen to twenty names bestowed on them, and one, now called *Brachionus capsuliflorus*, has in the course of time been referred to in literature by no fewer than forty-seven names! H. K. Harring, an American biologist, made an attempt as long ago as 1913 in his *Synopsis of the Rotatoria*, and in his later papers, to produce order out of chaos and bring to light the names which, by a strict application of the international rules of zoological nomenclature, should take priority.

Unfortunately, in spite of his efforts, nearly all writers have continued to use the incorrect names in most textbooks and popular works dealing with pond life. Some of the names used below, therefore, may be unfamiliar to readers who have become accustomed to the incorrect but more familiar names, such as *Melicerta* for the case-building rotifers and *Floscularia* for the 'Floscule'. For ease of reference, however, the familiar names are also given in the text.

At least 530 species of rotifers are found in the British Isles, so that any possibility of dealing adequately with each one, or of giving means of identification here, is ruled out. All that can be done is to mention very briefly a few representatives taken arbitrarily from each of the groups which have freshwater members.

Order MONOGONONTA (Gr. = one ovary)

Sub-order PLOIMA (Gr. *ploimos*, sea-worthy)   The members of this sub-order are mostly free-swimming rotifers, but some, and these are perhaps the most interesting, are parasites on algae or small invertebrates. *Proales parasita* and *Ascomorphella volvocicola* (Plate **20**), for instance, occur inside the spheres of *Volvox*, feeding apparently on the daughter cells. *Proales wernecki* is found on the alga *Vaucheria*, and causes an oval gall on the side of the filament. The eggs of the rotifer are laid inside the gall. *Proales gigantea* is parasitic in the egg-masses of the water-snail, *Lymnaea*. *Pleurotrocha daphnicola* (= *Proales daphnicola*) cements its eggs to the shells of *Daphnia*, and in the adult stage is usually attached by its 'toes' to the same place. Rotifers of this species hold the record for the greatest depth at which freshwater rotifers have been discovered, for specimens were found in an oligochaete worm dredged from a depth of 150 metres in Loch Ness.

Rhinoglena frontalis (= *Rhinops vitrea*) is a conical-shaped rotifer which is usually found during winter or spring months. It is remarkable in that the male, which is half the size of the female, is well developed and possesses a complete alimentary canal and a mastax. The many species of the genus *Brachionus* are,

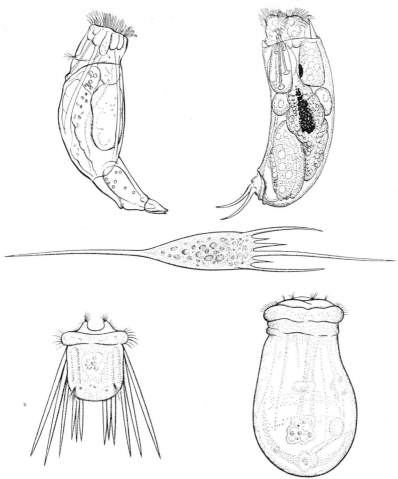

Fig. 43 Rotifers: *Above: (left) Pleurotricha daphnicola* (× 100);
(right) Trichocerca porcellus (× 500). *Middle: Kellicottia longispina* (× 100). *Below: (left) Polyarthra dolichoptera* (× 250);
(right) Asplanchna priodonta (× 75)

perhaps, the best known of all the rotifers, being common and large in size. They
are shaped rather like a wine-glass without the base, and the width of the trunk
or middle portion makes them particularly suitable subjects for examination of
their internal organs under the microscope. The females attach their eggs to
the rear of the lorica and carry them about until they hatch. (Plate **18,** 2) *B. caly-
ciflorus* is often found in lakes, ponds and canals rich in phytoplankton. *B. rubens*
attaches itself in numbers to *Daphnia.*

The various species of the genus *Keratella* (Plate **20,** 3) have loricas somewhat similar in shape to those of *Brachionus*, but with a pattern of polygonal facets on the dorsal side.*Kellicottia longispina* is easily recognized by the very long spine at the rear end of the body and the six smaller spines of various sizes pointing forwards at the anterior end. It is a characteristic species of lake plankton. *Polyarthra dolichoptera*, another species found in open water, has twelve long spines which, although attached to the fore-part of the lorica, point backwards, but are capable of some movement and are possibly useful for defence. A large and common urn-shaped rotifer commonly found in lake plankton is *Asplanchna priodonta*.

Trichocerca porcellus (Plate **20**, 8), found in weedy ponds and ditches, has been well described as 'like a plump, well-filled sausage'. It is frequently taken amongst masses of filamentous algae, but swims readily with a rolling motion.

Synchaeta pectinata (Plate **20**, 1) is a rotifer of the open water and occurs in the surface plankton. It is easily recognized by the bluish colour of its eye, which contrasts with the reddish hue usual in rotifers.

When examining the common freshwater louse, *Asellus*, one frequently finds rotifers attached to the underside of the crustacean, usually near the constantly moving appendages on the abdomen. These are generally specimens of *Testudinella elliptica*, or *T. truncata*, which derive considerable benefit in the form

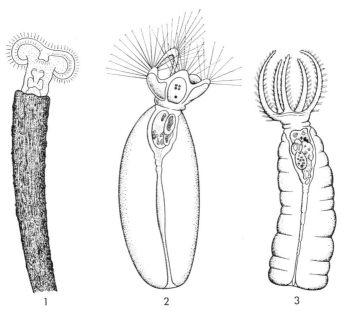

1   2   3

Fig. 44   Fixed rotifers: 1 *Limnias melicerta* (× 50).
2 *Collotheca campanulata* (× 120). 3 *Stephanoceros fimbriatus* (× 50)

of food particles and a fresh supply of oxygen from their proximity to these current-producing structures.

Sub-order FLOSCULARIACEA (L. *flosculus*, a flower)   To this order belong two of the best-known genera of rotifers, both of which are favourite objects for the microscopist. *Floscularia ringens*, under its old name *Melicerta ringens*, has been figured or described in almost every book dealing with the natural history of fresh water which has been published in the last hundred years, and it is particularly unfortunate that by an inexcusable mistake the wrong name was applied to the creature.

The delicate body of *Floscularia ringens* is encased in a cylindrical tube built by the creature from 'bricks' made from particles in the water and its own excrement. The creature is found attached to water-plants and particularly to the Canadian water-weed, *Elodea*, the moss, *Fontinalis*, and filamentous algae, and is just visible to the naked eye as a little brown spot on the green leaf. After a disturbance, the little tube is all that will be seen under the microscope, but in a short time the four lobes of the coronal disc will be carefully protruded at the top of the tube (Plate 18, 4), and soon the flow of particles in the water, caused by the wheel-like movements of the cilia, will be noticed.

The 'bricks' are made in a glandular organ below the mouth, particles being churned round and round and becoming cemented into pellets the shape of which varies somewhat. They are then deposited in orderly fashion around the top of the case. It is interesting to place in the water in which one of these rotifers is being observed some particles of a colouring substance such as carmine, when the complex currents caused by the cilia can be studied and the process of pellet-making can be watched.

Limnias ceratophylli might at first glance be mistaken for a small specimen of *Floscularia*; its case, however, is not made of pellets, but of a horny secretion of the body. In time the case may become covered with adhering particles of matter and be difficult to see. The crown of cilia consists of only two lobes. This creature seems to favour water milfoil, *Myriophyllum*, and hornwort, *Ceratophyllum*, but does also occur on other plants.

Conochilus is a colonial species, with the individuals attached in a common ball of mucus. The colony, which may be up to 1 mm across, is free-swimming and is found not uncommonly in large ponds and lakes. *C. hippocrepis* and *C. unicornis* are the two British species.

Sub-order COLLOTHECACEA (L. *collum*, neck; *theca*, a case)   The members of this order are, like those of the last, fixed and tube-dwelling forms. *Collotheca campanulata*, which is the name which must now be given in place of *Floscularia campanulata*, is a truly lovely creature, particularly when it is seen by dark-ground illumination under the microscope. It is found attached to the leaves or stems of water-plants, including the strands of filamentous algae. The body of the creature is enveloped in a wide transparent sheath, which is not always easy to see. When the creature is first located in the microscope field, the body will probably be contracted into the sheath, but when it eventually extends it will be seen to consist of an open funnel-shaped structure, around the rim of which are five protuberances, each bearing extremely long hair-like cilia. Unlike the cilia of other rotifers, these are not in constant motion, but are normally straight and comparatively still. When minute organisms come within reach of the funnel opening, however, they are inclined inwards to act as a trap. The food is taken into the funnel, and any unsuitable matter there rejected, the rest passing down to the mouth and so to the jaws.

The oval eggs, numbering usually about half a dozen, may sometimes be seen inside the sheath near its base, and frequently the developing embryo will be observed wriggling inside. The creature which hatches from the egg is similar to its parent, but with a crown of fine cilia, and, after breaking through the sheath, swims round before settling on a support.

Stephanoceros fimbriatus (= *S. eichornii*) is undoubtedly the most exquisite of all the rotifers and a great prize for microscopists. It is by no means a rare creature, and where it is found it usually occurs in fair numbers, often in my experience on milfoil, although *Elodea* is also a favourite plant. Like *Collotheca*, it is enveloped in a transparent sheath, but apparently of a more solid nature and bearing wide folds. The corona is drawn out into five long, pointed arms, which curve inwards and bear whorls of cilia on their inner sides. To see the arms extended and the cilia waving about bringing food particles to the gullet is indeed a beautiful sight. (Plate **18**, 3)

The males of both *Collotheca* and *Stephanoceros* are, like most male rotifers, small, insignificant creatures, and it is the females only which arouse our admiration.

Order DIGONONTA (Gk. = two ovaries)

Sub-order BDELLOIDA (Gr. *bdella*, a leech)   The rotifers in this sub-order owe their name to the way in which they creep about in a looping manner rather like a leech. Many are dwellers in mossy places and in temporary bodies of water such as roof gutters, and have remarkable powers of enduring in an encysted state in the conditions of frequent drought or extremities of temperature which can happen in these unusual surroundings. Students of this group have been able to store the encysted creatures in dried moss until required, when, after a short immersion in water, they once again become active rotifers.

Sometimes in a dried-up gutter or bird-bath a reddish sediment will be seen and, if a little of this is lifted out on the blade of a penknife and placed in water, specimens of a large bdelloid, *Philodina roseola*, will soon be swimming about. This was the rotifer to which Baker was referring in the letter of which extracts were given on page 104. It is sometimes found in association with a minute alga,

Sphaerella lacustris, which favours similar locations to the rotifer and which also has a reddish colour. (Plates **18, 1; 20, 2**)

Philodina citrina is often found crawling about on Polyzoa, particularly near the area bearing their crown of tentacles.

Rotaria rotatoria (= *Rotifer vulgaris*) is one of the commonest rotifers found in ponds. It has a long, cylindrical body tapering to the rear end. *Rotaria neptunia* is an exceptionally long rotifer with two curious spurs about the middle of the foot portion and three long toes at its extremity. Its great length gives it an ungainly appearance as it crawls about in the water. (Plate **20, 6**)

Embata parasitica (= *Callidina parasitica*) is usually attached to the appendages of crustacea, such as the freshwater shrimp, *Gammarus*, and the freshwater louse, *Asellus*. The young are born alive.

Class *GASTROTRICHA*

(Gr. *gastr-*, stomach; *trich-*, hair)

The minute creatures called 'hairy backs' or 'bristle backs' have features in common with the rotifers, and are usually considered with them, although now placed in a separate class.

They are found in aquatic debris and among filamentous algae, swimming by means of the cilia with which their bodies are so liberally covered. Although not rare creatures, they are generally overlooked on account of their size, for most of them are smaller than Protozoa, the largest being only 0·5 mm long.

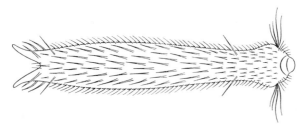

Fig. 45 Hairy back, *Chaetonotus* (× 400)

Their bodies are almost cylindrical, although flattened a little on the underside. The skin of the back bears scales in addition to the cilia and the long hairs which have given rise to the creatures' names.

The mouth, situated at the head end, leads to a simple alimentary canal, which ends in an anus opening on the back. The food is believed to be mainly organic debris, but specimens of Gastrotricha have been seen devouring Protozoa.

These creatures are hermaphrodite and the eggs are fastened to plants and other submerged objects.

At least seven species have been found in Britain, of which those belonging to the genus *Chaetonotus* are the best known.

Phylum ACANTHOCEPHALA
Proboscis Roundworms

Sometimes, when examining a freshwater shrimp, *Gammarus*, or freshwater louse, *Asellus*, an animal, roughly oval in shape and reddish-brown in colour, may be seen lying inside the crustacean just below the dorsal surface. This is the larva of a proboscis worm, *Echinorhynchus*, of which there are several species.

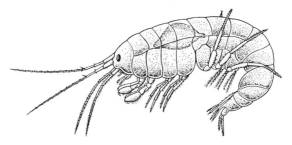

Fig. 46   Larva of proboscis roundworm, *Echinorhynchus*, in a freshwater shrimp

The acanthocephalan worms, although placed in a separate phylum are, believed to have affinities with the roundworms. They are parasites, having no alimentary canal and absorbing their food from the hosts through their body surface. At the front end is a proboscis, provided with hooks, with which the animal attaches itself to its prey. The freshwater species are only about 2–3 mm long and pass their larval stages in arthropods and the final stage in fish. *Neoechinorhynchus* passes its larval stage in the alder-fly larva, *Sialis*, that of its adult stage in fish.

I O: MOSS ANIMALS

Possibly the most beautiful and interesting of all freshwater invertebrates are the moss animals, yet even their existence is unknown except to those who have taken more than a passing interest in freshwater biology.

The old name for the group, Bryozoa (Gr. *bryon*, a lichen; *zöon*, an animal), refers to the mode of growth of these animals; many individuals, or zooids as they are called, are joined together so that they often resemble a patch of moss on stones or tree-roots. Sometimes these colonies form huge masses. In some species the colony can move along as a unit.

The majority of moss animals live in the sea. The freshwater species, although not numerous, are commoner than is generally realized, but they are usually overlooked except by those searching particularly for them.

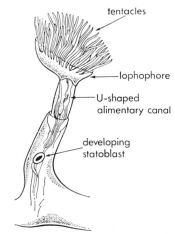

Fig. 47 Structure of a moss animal (× 15)

The classification of these interesting creatures has given systematists considerable trouble. The early naturalists thought they were plants and later two names, Polyzoa (= many animals) and Bryozoa, were independently created for this group of animals. Under present classification, two new phyla take the place of the older Bryozoa or Polyzoa: ENDOPROCTA (Gr. *endon*, within; *prōktos*, the anus) and ECTOPROCTA (Gr. *ektos*, outside), the names distinguishing between the species in which the anus opening is inside or outside the circle of tentacles. The Ectoprocta also have a well-developed body cavity, which is lacking in the Endoprocta. The freshwater species are included in the Phylum Ectoprocta and the class Phylactolaemata.

The general structure of a typical moss animal (Fig. 47) comprises a transparent sac forming the body, which in some species is covered with a brown opaque sheath and in others is immersed in a jelly-like substance. This contains a U-shaped alimentary canal with, of course, a mouth at one end and an anus at the other. There is no heart, nor are there blood-vessels.

The most noticeable feature of the animal is the crest of tentacles which is mounted on a kind of platform called the *lophophore*. A powerful muscular system enables the lophophore and its crown of tentacles to be withdrawn into the body or protruded at will.

There can be few more lovely sights, particularly when seen under darkground illumination of a low-power microscope, than a colony of these creatures with their delicate, translucent, flower-like crowns of tentacles extended and waving freely.

The tentacles are covered with fine hair-like cilia, and the rhythmic movement, which in some species has been likened to the waving of corn, sets up currents in the water to bring to the mouth (situated beneath the lophophore) the minute organisms on which the creatures live. Diatoms, infusoria, rotifers and similar fare caught in the whirlpool are swept down the alimentary canal into the stomach.

The multiplication of the moss animals is effected in three ways: by sexual reproduction, each animal being hermaphrodite, by budding and by the production of statoblasts, which can be considered as a kind of bud produced internally.

Sexual reproduction has been studied in some species. In *Cristatella*, for example, the ovaries which are found just inside the common wall of the colony each mature a single egg, which is fertilized and develops *in situ*. In due course the free larva escapes from the colony through the opening caused when one of the nearby full-grown animals dies and decays.

Budding takes place round the edges of colonies, and in time huge masses are formed.

Perhaps the most interesting and certainly the most easily studied method of reproduction is the production of *statoblasts* internally. These seed-like objects consist of groups of cells enclosed in a tough, resistant wall, arranged in the form of two flat saucer-shaped plates stuck together at their edges. The statoblasts leave the body of the animals when the colony disintegrates, usually at the end of summer, and remain dormant until spring, when the two plates of which the statoblast is formed come apart. The young animal, a miniature replica of the adult, emerges, floats about passively at first until it finds a resting-place, where it settles and starts a new colony. Except in *Cristatella* and to a lesser extent *Lophopus*, which possess the power of movement in the colonial state, the freshwater moss animals live a sedentary existence when adult.

The statoblasts of the different species vary in appearance, and are sometimes useful as a means of identification. No statoblasts, however, have been found in the case of *Paludicella*.

Close examination of the debris floating at the edge of ponds and lakes in spring will often reveal the presence of many statoblasts, and this is usually a good way of determining whether or not bryozoans are present in that particular stretch of water. If the statoblasts are kept under observation in a small receptacle, the development of the animal may be studied.

In floating at the surface of the water the statoblasts, of course, are subjected to all the rigours of winter, and must often be frozen for long periods. There is some evidence that their germinating power is increased by a certain amount of freezing, and in experiments which were carried out, one group of statoblasts placed in water not subjected to frost even failed to develop, whereas a similar group which was frozen developed normally when afterwards exposed to suitable conditions. It seems probable that some enforced period of rest or absence of development is necessary before the statoblasts achieve their optimum germination.

Statoblasts must often adhere to the bodies of waterside creatures, or become embedded in the mud clinging to the feet and legs of water-birds, and no doubt are thus carried from one water to another.

In *Paludicella* and *Victorella*, both of which belong to a different group from the remaining moss animals described in this chapter, there are specially modified external buds which persist through the winter after the rest of the colony has died down, and in spring these split open and a young organism emerges. These external buds are referred to as 'hibernacula'.

Bryozoan colonies are usually found adhering to submerged objects in fairly deep water. Decaying pieces of wood, rotting branches, the submerged rootlets of alder and willow trees, are likely places to search for them, while others are to be found on the underside of water-lily leaves, on the stems of water-plants and amongst the fronds of duckweed. Locations that stand out in my memory as having produced good 'hauls' of these interesting creatures are boating-lakes and disused portions of canals, and I think most experienced pond-hunters would endorse this fact. The submerged woodwork of the landing-stages of the former and the sides of bridges of the latter rarely fail to produce specimens. A net having a scraping edge, such as is illustrated on Plate 55, A, is particularly suitable for collecting bryozoans from such places.

Gentle movements or circulation of the water in which they are living seems to have a beneficial and stimulating effect on the creatures, and this may explain the choice of such places as boating-lakes and canals, which on first thought might be considered too turbulent a position in which to find creatures so sensitive to disturbance as they are.

When located, bryozoans will rarely, if ever, be found with their tentacles extended, for of course violent disturbance of their habitat causes them instantly to retract, and all that will be seen will be blobs of jelly-like material. The long rope-like gelatinous colonies of *Cristatella* may then easily be mistaken for the spawn of water-snails, and the branching colonies of *Plumatella* and *Fredericella* often resemble, both in colour and shape, the tree rootlets on which they are so frequently found. *Paludicella* can be found forming soft spongy masses on the bank of a lake which has been undercut by the current. To the touch they resemble the bodies of the freshwater sponges with which they are often intermingled, or a soft mossy growth, and it was, of course, this latter similarity that gave rise to the names Bryozoa or moss animals, as mentioned earlier in this chapter.

Some species, particularly *Plumatella* and *Fredericella*, frequently cause trouble at waterworks. When alive, the colonies encrust the interior of the water-pipes, thus restricting the flow, and their dead bodies on decay may give a noticeable taste or odour to the water supply.

Summer, late summer especially, is the time to find bryozoans. By June the colonies are of a readily detectable size and, of course, as the summer advances they grow bigger. Normally the colonies die down about October, although exceptionally they may be found right into the winter, and colonies of *Lophopus* have, in fact, been known to survive a winter, including a period of frost in an open pond.

In late summer and autumn the colonies contain many statoblasts, and it is worth while collecting at this stage and keeping the colonies in a micro-aquarium in order to study the liberation of the statoblasts and their subsequent development.

In the pages which follow, the commoner British species are described in some detail, but the freshwater Ectoprocta have not received a great deal of study, and it may well be that forms such as *Urnatella*, *Pottsiella* and *Pectinatella*, not so far recorded in Britain, will be found in this country, and will have to be added to the list of British species.

Genus *Cristatella* The only species, *C. mucedo*, is usually found in colonies about 50 mm or more in length and 6 mm wide which resemble the egg cluster of the pond snail, *Lymnaea stagnalis*. They are found most frequently on the undersides of water-lily leaves exposed to full sunlight, or on the stems of this plant and others such as *Elodea*.

The underside of the colony is rather like the 'foot' of a snail, and enables it not only to adhere to surfaces but also to creep about from place to place. In specimens which were observed, a speed of about 25 mm per hour was recorded. How this movement is achieved is not quite clear, but is probably through the individual animals of the colony reacting in the same way to an external stimulus such as sunlight, thus providing a unified effort which results in the entire colony moving.

The individual animals bear a horseshoe-shaped crest of up to eighty or ninety tentacles, thus achieving the distinction of possessing more tentacles than any other British moss animal.

As if proud of its crowning glory, *Cristatella* is more ready to display its extended tentacles than the others. Not only is a good deal of rough treatment needed to cause them to be withdrawn, but once they have been retracted they are very quickly extended again. *Plumatella, Fredericella* and *Paludicella*, on the other hand, retract at the slightest disturbance, and it may be an hour or more before they will again extend their tentacles.

The circular statoblasts of *Cristatella* are about 1 mm in diameter, dark brown in colour and have round their circumference a number of spines each terminating in two or more hooks. When the statoblasts are eventually liberated from the decaying body of the animal, they become attached by these hooks to water-plants and other objects, and are thus carried about. (Plate **21,** 2)

Genus *Lophopus* It is of interest that *Lophopus* was the first moss animal to be described, both in this country and abroad. Trembley, the Dutch naturalist, discovered it in waters near The Hague in 1741, and Baker in a book published in 1753 described its occurrence in this country and named it the 'bell flower animal'. (Plate **19,** 2)

Although the individual animals are larger than *Cristatella*, the colonies are much smaller (about 12 mm across), and rarely have more than about a dozen organisms in them. They are enclosed in a transparent jelly-like covering, which becomes deeply cleft or lobed as the colony grows, and it is at these points that the colony eventually divides. The rootlets of duckweed and the stems of *Elodea* are the most likely places to find *Lophopus*, but the tiny blobs of jelly are not easily seen.

The colonies have a disc on their undersides, which serves as a foot for attachment. Although *Lophopus* is not given to creeping about when adult, small, slight movements have been observed in the earlier stages of its development. One colony which was studied shortly after dividing from the main colony moved nearly 25 mm in three days. The colonies, however, soon become detached from their supports, and settle down some distance from their original position, which would, of course, give grounds for thinking a creeping movement had taken place unless great care had been taken in observation.

Like *Cristatella*, *Lophopus* is not unduly sensitive and soon extends its crown of about sixty tentacles after being disturbed.

The statoblasts are oval in general shape, but are pointed at both their narrower ends. They are brown in colour and reticulated, but bear no hooks. (Plate **21,** 2)

Reproduction by sexual means is sometimes observed, the larval form being light in colour when it escapes from the colony; it floats about in the water until settling on a suitable support.

Genus *Plumatella* Considerable confusion exists even today regarding the classification of the genus *Plumatella*. Some authorities consider there are eight British species, whereas others recognize only two. As the modern tendency is to increase rather than decrease the number of species in all branches of biology, the distinguishing features of each of the eight so-called species are given below. Judged purely on the shape of the statoblasts, it is possible to differentiate between two distinct groups—those with statoblasts that are nearly round and those with oval-shaped statoblasts. The different forms in these two main groups —call them species or varieties as you will—vary in their mode of growth.

I Statoblasts nearly round:
P. repens, branches spreading out.
P. coralloides, branches coalescing.
P. fungosa, spongy tubes tightly packed together.
P. flabellum, another variant of the spongy type.
II Statoblasts markedly longer than broad:
P. emarginata, branches spreading out.
P. muscosa, branches coalescing.
P. benedeni, P. spongiosa, spongy.
P. fruticosa, spreading branches, very long statoblasts.

Formerly a separate genus, *Alcyonella,* was introduced to include the spongy forms, but this genus is not accepted by present-day taxonomists, since both of the types given above may grow, not only in clearly defined ramifying or spongy forms, but also in intermediate forms. It seems logical to consider the spongy types to be merely varieties of the main species. Another specific name, *P. jugalis,* was given to colonies derived from larvae as opposed to those arising from statoblasts, but as it is probable that any species of *Plumatella* can be found in *jugalis* form, this name also has been abandoned.

From the intricacies and uncertainties of classification, it is a relief to turn to a discussion of the living animal.

Plumatella is found attached to the underside of water-lily leaves, stems, stones, tree rootlets and other submerged objects not exposed to full sunlight, particularly favoured spots being under canal bridges and boat landing-stages. The ramifying type of colonies adhere closely along the whole length of their branches to their support, sometimes covering an area of as much as 80 cm^2. The spongy type form bushy masses of closely packed tubes and, as the new season's tubes grow on the old dead tubes, huge colonies perhaps 75 mm thick and 30 cm or more in length may be found.

The lophophore is horseshoe-shaped, and bears about fifty tentacles, which are held more erect than those of *Cristatella*.

The statoblasts are reticulated but without spines, and towards the end of summer are found in great numbers within the body of the creatures. (Plates **19, 1; 21, 3**)

Plumatella is a very sensitive animal, and when it is disturbed it may be half an hour or more before the tentacles are extended again.

Genus *Fredericella*   *F. sultana* is the only British species and, although widely distributed, it readily escapes detection, partly on account of the smaller size of the colonies compared with the other moss animals, and also because the colonies are usually covered with algae or sand particles, and thus closely resemble the surroundings, particularly if these are tree rootlets. As often as not *Fredericella* will be found by accident.

It occurs in shady places in the form of branching tufts hanging freely from the stems of water-plants, the rootlets of alder and willow trees and also on stones.

The corona is quite different from the preceding species, being almost circular and containing only about twenty-four long slender tentacles, held erect when extended.

The statoblasts are bean-shaped without any spines or reticulations, but are not produced in such numbers as in the previous species.

Genus *Paludicella*   *P. articulata* is the only species, and the characteristic branches of club-shaped cells joined end to end will readily distinguish it from the other genera. It is found in loose, mossy bunches in shady places, such as under canal bridges or under banks which have been undercut by the passage of boats, attached to stones or tree rootlets, part of the colony being fixed to the support and the rest free.

The corona is circular, but there are fewer tentacles than in *Fredericella*, and never more than sixteen, which are held up straight when extended.

Paludicella is the shyest of all the freshwater moss animals, and may take several hours to show its tentacles after being disturbed.

Statoblasts have not been observed in this species, and reproduction is presumably by the production of true eggs as well as by budding.

The muscles for retracting the tentacles can be particularly well seen in *Paludicella*, especially under a microscope fitted with polarizing equipment.

Genus *Victorella*   For the sake of completeness brief mention must be made of this moss animal, for although it is essentially an inhabitant of brackish water it may stray into fresh water near the coast.

It was named after the Victoria Docks in London, where it was first found. *V. pavida* is the only British species, and has since been found in other British localities.

The colonies consist of a main stem, branching off from which are the

I

individual 'heads'. From these, either new stems or further heads may originate. There are no statoblasts.

The study of these beautiful creatures is a most fascinating pursuit. The most convenient way to have specimens always available for examination is to keep them in small aquaria, and they can be kept quite successfully in any small receptacle providing they are not crowded. The water should be shallow, and the surface area great compared with the depth in order that the water may be well oxygenated. Due regard should be paid to the likes and dislikes of the various species regarding light.

The easiest way of providing food is to fill the receptacle in which they are kept with some of the water from which they were taken, or from similar locations, and to keep adding regular supplies of it.

Artificial circulation from a small aquarium air pump is helpful, and the creatures seem to be stimulated by a constant dripping of water if this can be arranged from a nearby tap.

I I: THE MOLLUSCS

Only two classes of the large phylum Mollusca (L. *molluscus*, soft) are represented in fresh water—Gastropoda (Gr. *gaster*, the belly; *podos*, a foot), the limpets and snails, and Bivalvia, in which are included the mussels and cockles. Molluscs of one kind or another are among the first creatures to be encountered by students of freshwater life, and they are found all the year round, although in the coldest weather they usually bury themselves in the bottom mud.

The shell, which is such a characteristic feature of all of them, renders confusion with any other creatures impossible, and its type also serves to distinguish between the members of the two classes, for while that of a lamellibranch consists of two parts or *valves* which are hinged together, the gastropod shell is all in one piece. On account of these shell differences the animals are referred to as bivalves and univalves, respectively. In the past the study of shells alone led to numerous errors in describing as new species the very many local varieties arising from differences in their habitats, but in recent years there has been a welcome change in emphasis and a good deal of research has been directed to the distribution, ecology and general biology of the living animals.

Molluscs have soft bodies with no internal skeleton, and without any division into segments. They do not possess an abundance of appendages as do arthropods. Very characteristic of them is the *foot*, a muscular development of the ventral surface of the body. In snails the foot forms a flat sole on which they creep along, lubricated by a trail of slime secreted by the glands at the front of the foot. In mussels it is a triangular structure which can be protruded from the front end of the shell to serve as an anchor or to plough a way along the surface of the mud on which they live.

A structure of great interest in molluscs is the *mantle*, a fold of the body-wall which in snails covers much of the body and in mussels forms two flaps or lobes, one on each side of the body. It is the mantle which secretes the substances of which the shell is composed. The shell consists largely of calcium carbonate or chalk, and is made up of three layers each differing from the others in composition. The three layers can easily be seen by carefully examining an empty shell, preferably of a mussel. The thin, outer skin-like layer which serves as a protection to the shell, particularly in the case of aquatic species where it guards against the action of organic acids in the water, is composed of almost pure *conchyolin*, a substance somewhat similar to the chitin of insects. In older shells much of this skin may have been rubbed off to reveal the middle layer which forms the major part and consists largely of calcium carbonate crystals arranged perpendicularly to the surface of the shell. The inner lining of *nacre* or mother-of-pearl is also made of calcium carbonate, but in this case the crystals are lying parallel to the shell surface.

The nacre is secreted by the general surface of the mantle, but the two outer layers are provided by the fringes of the mantle, and as the animal grows the shell is extended at the edges, new whorls being added in the case of snails. The rate of growth is not uniform and thus the additions to the shell appear as lines parallel to the edges, and after some experience these can be used as an indication of the age of the inmate.

It is believed that, whereas most of the smaller snails and the cockles have a short life of little more than a year, the freshwater winkles, *Viviparus*, and the larger species of ram's-horn snails, *Planorbis*, live at least two or three years, and the mussels are very long-lived, ten or fifteen years being a common age.

The large amount of calcium which enters into the composition of the shells of molluscs makes it necessary for them to live in places where there is an adequate supply of calcium salts, and with decreasing amounts of calcium in the water the number of snails becomes progressively fewer. On the other hand, they vary greatly among themselves in their requirements of calcium, and it is possible to group the freshwater molluscs into those that need and are found only in 'hard' waters—the so-called *calciphile* species (L. *calcis*, lime, chalk; Gr. *philos*, fond of)—and those that can thrive in softer waters. It is notable that specimens found in the latter are on the whole smaller in size. It should be remembered that ponds and streams not in chalky or limestone districts may still be rich in calcium on account of the drainage from agricultural land which has been well limed.

Quite apart from the needs of shell-making, calcium is important to freshwater molluscs in other ways. Firstly, the greater profusion of both higher plants and algae in waters of high calcium concentration provides more food for them. Then, as every gardener knows, lime has the property of agglutinating particles of clay. Such particles in suspension in the water are thus precipitated and the water is kept from becoming turbid, a condition of importance to those molluscs that breathe by means of gills.

It is perhaps worth noting in passing that the shell of a mollusc is firmly attached to the body with powerful muscles, and must be regarded as part of the animal and not merely an external shelter, as is, for instance, the larval case of a caddis.

All molluscs have a body-cavity distinct from the food canal, a heart and blood. It will be as well, however, to discuss the more detailed characteristics of gastropods and lamellibranchs separately, as the two classes differ greatly in both their structure and their mode of life.

The Water-snails and Freshwater Limpets

Class *GASTROPODA*

Thirty-six species of freshwater gastropods are found in Britain, occurring in a wide variety of environments ranging from rapid mountain streams to marshy areas on land. They are active, if slow-moving, creatures, in contrast to the mussels which remain stationary for long periods, and they glide about over plants, stones or even the underside of the surface film of the water in search of the algae or higher plant tissues on which they mainly feed. From the fact that the front part of the body carries a mouth and a pair of sensory tentacles with eyes at their base, it is given the courtesy title of head, although there is, in fact, no distinct head. The major part of the body is always inside the shell, and by the contraction of the muscles attaching the body to the shell, the rest of it—the head and foot—can also quickly be withdrawn inside. The body is twisted over to one side or the other depending on the species, and the internal organs are all

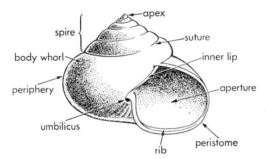

Fig. 48 Shell structure of a univalve

massed together into this enclosed part of the body, which is therefore called the *visceral hump*. The spiral nature of the shell, which is so characteristic of snails, is of course a result of the twisting of the body. If, when one is looking into the opening of a shell held so that its apex or tip is pointing upwards, the opening is to the observer's right-hand side, the shell is said to be *dextral*; if to the left, *sinistral*. Most species of snails are dextral, although sinistral coiling is the rule in other species. Even in normally dextral species, individuals with left-handed twists sometimes occur.

A typical snail shell (Fig. 48) can be considered as a long conical tube twisted into a series of coils, each of which is called a *whorl*. The largest whorl is the *body-whorl* and the remainder are called collectively the *spire*. The line separating two whorls is called a *suture*. The first whorl of the shell at the apex was the part originally occupied when the snail was small, and the succeeding whorls have been added to keep pace with the growth of the body.

To see the mouth of a water-snail such as the great pond snail, a specimen should be watched closely as it creeps over the glass front of an aquarium. At intervals the mouth aperture will be seen to open and close, and reveal the rasp-like tongue or *radula* with which algae and higher plant tissues are scraped off. The radula bears many rows of fine teeth and, since their form and arrangement vary among the species, these provide a means of identification.

Freshwater gastropods belong to two sub-classes Prosobranchia (gill-breathers) and Pulmonata (lung-breathers). The third sub-class of Gastropoda, Opistho-branchia, is exclusively marine except for the family Succineidae, two or three species of which live on the aerial parts of emergent water-plants and sometimes fall into a net when one is pond-hunting.

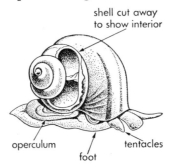

shell cut away
to show interior

operculum   tentacles
foot

Fig. 49   Structure of an operculate mollusc

The most obvious characteristic of freshwater members of the first group is their possession of either a horny or chalky plate called the *operculum* attached to the foot, from which feature they are called operculates. When they withdraw into their shell, the operculum fits snugly over the opening and closes it. This structure must not be confused with the layer of hardened mucus with which some land-snails close the aperture of their shells when hibernating or indulging in periods of rest. Freshwater operculates breathe by means of a gill attached to the mantle, and live mainly in well-oxygenated and clear waters, such as streams and rivers. Except in the family Valvatidae, both male and female individuals exist, and in some species the young are born alive.

Operculates

The nerite[*][1] *Theodoxus fluviatilis* (= *Neritina fluviatilis*) has an attractive almost globular shell about 7–10 mm high, varying greatly in colour from yellow through brown to black, but always bearing irregular pink, white or purple blotches. The nerite is found in most parts of England, mainly in running water and the margins of lakes attached to stones or plants. The operculum is reddish in colour. (Plate **24, 8**)

[1] In the descriptions which follow in the rest of the chapter, species peculiar to 'hard' waters are indicated thus: * and those which are found in quite soft waters thus: †. Those not indicated can live in water of a moderate degree of 'hardness'.

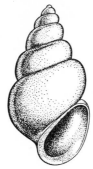

Fig. 50 Operculates: *Left:* nerite, *Theodoxus fluviatilis,* 7–10 mm high. *Middle:* Jenkins' spire shell, *Potamopyrgus jenkinsi,* 4–5 mm high. *Right:* valve snail, *Valvata piscinalis,* 5–7 mm high

Jenkins' spire shell†, *Potamopyrgus jenkinsi* (= *Hydrobia jenkinsi*), a small dark snail, about 4–5 mm in height, is of interest because it was, until about the end of the nineteenth century, confined to brackish water, but has now spread in an amazing manner to freshwater habitats all over the British Isles, and is found abundantly in running water in most parts of the country.

Bithynia tentaculata★ is common in most parts of the country on stones, weeds and logs, in larger stretches of still water and occasionally in streams, but never in small ponds. *B. leachii*★ is not so plentiful, and is found mostly in the south and east of England, usually in slowly running water or larger lakes. They are both small snails, *B. tentaculata* having a shell about 15 mm in height whereas that of *B. leachii* is only about 6 mm. Their opercula are chalky and brittle, and the tentacles of both the male and female are long and slender. The eggs, which are enclosed in a gelatinous capsule, are deposited on stones.

The best-known freshwater operculates are the freshwater winkles belonging to the genus *Viviparus.* Two species, *Viviparus viviparus*★ (= *Paludina vivipara*) (Plate **22**, 1, 2) and *V. fasciatus*★ (= *P. contecta*), are found in slowly running water in many districts as far north as Yorkshire. Occasionally they will be found in ponds, but probably only because they have been introduced there. They are both large snails with shells sometimes over 35 mm in height and of a greenish-brown colour with three brown spiral bands running around each whorl. The operculum is hard and horny. The male is larger than the female, and may be distinguished by his short, blunt right-hand tentacle, both of the female's tentacles being long and tapering. About fifty eggs are laid at a time, and are retained within the body of the female until the development of the embryos is complete, when the young snails are released into the water fully formed.

The valve-snails, *Valvata*, of which there are three species, are easily recognized by the snout-like appendage on the head between the tentacles; a further distinguishing feature is the division of the foot in front into two lobes. When the animals are crawling about actively, two structures, a feathery gill and a slender filament also for respiratory purposes, are protruded from under the shell on the right-hand side. Unlike most operculates the valve-snails are hermaphrodite. The eggs are laid in flat gelatinous masses.

Valvata piscinalis† is the commonest species, occurring in running water in most parts of the country. The shell is about 6 mm in height, but somewhat flattened with a definite spire. *V. cristata†* has a flat coiled shell about 4 mm in diameter, like that of a small ram's-horn snail, but easily distinguished by the presence of the operculum, a feature not, of course, found in a pulmonate. It inhabits muddy streams where it is usually found among the water-plants.

V. macrostoma*, a species with a shell about the same in diameter as *V. cristata*, but with a definite spire, occurs in marshes and ditches in southern and eastern England.

Pulmonates

The freshwater pulmonates have no operculum and no gills. Although they are able to extract a certain amount of dissolved air by simple diffusion through any of their tissues in contact with the water, they obtain their main supply of air by taking it in at the surface of the water into a cavity, the lung, between the mantle and the dorsal wall of the body. The opening to this cavity may be seen as a small, round, tubular structure on the right-hand side of a pond-snail when it is at the surface of the water in an aquarium. The area of the mantle surrounding the air-cavity is well supplied with blood, and the aerated blood passes to the heart, whence it is pumped to all parts of the body.

We can have little doubt from this method of breathing that those pulmonates now living in fresh water are descended from land-snails which returned to the water. Nevertheless, in well-oxygenated waters they can remain submerged for long periods without rising to the surface for more air.

When a pond-snail is renewing its air supply at the surface, the end of the respiratory aperture is brought into communication with the atmospheric air and, by muscular movements of the dorsal wall of the body, air is taken into the mantle cavity. Then the aperture is closed and the snail descends below. The store of air gives the snail buoyancy, and occasionally a specimen down in the water will be seen to let go of its support and float rapidly to the surface. On the other hand, if disturbed it will expel the air, withdraw quickly into the shell and sink like a stone.

Each individual pulmonate snail contains both male and female organs, and all therefore are capable of laying eggs. Although self-fertilization does occur in some species, the union of two individuals is the more usual method of reproduction.

This group contains the pond snails proper, which belong mainly to the genus *Lymnaea*. They have thin shells, usually brownish in colour, mostly conical in shape, with a sharply pointed spire. Their tentacles, which are not retractile, are flat and triangular in shape, each having an eye at its base.

The great pond snail*, *Lymnaea stagnalis*, is common in larger ponds throughout the country, and is also sometimes found in slowly flowing rivers and canals. The shell may attain a height of 55 mm, so that it can hardly be mistaken for any other species. The eggs are laid in sausage-shaped gelatinous masses on the leaves and stems of plants, the underside of water-lily leaves being a particularly favoured place. The great pond snail, unlike most of its kind, is not solely vegetarian, but will readily devour decaying animal matter, and has even been known to attack living animals such as newts and small fish. (Plates **22**, **4**; **23**, **1**)

A contrast in size with *L. stagnalis* is the dwarf pond snail†, *L. truncatula*, for its shell is about 12 mm high. *L. truncatula* is a marsh species living mainly out of the water, and is very abundant in flooded or swampy pastures. Its role as the normal intermediate host of the liver fluke of sheep (see p. 97) has made the life-history of this species of more than usual interest. The eggs, about ten in number, are deposited in April in round gelatinous masses on mud or dead leaves. They hatch in three weeks, and by July the young snails are themselves ready to lay eggs. There are thus two and possibly three generations of the snail in a season.

The marsh snail†, *L. palustris*, inhabits not only marshy areas, but also ponds, ditches and the margins of rivers and lakes; although widely distributed, it is not abundant. The shell is somewhat elongated and about 20 mm high.

Undoubtedly the commonest of all our pond snails is the wandering snail†, *L. peregra*, which is found in all types of freshwater habitat and even in brackish water. The shell is very variable in form and size, but is usually about 18 mm in height. The eared pond snail*, *L. auricularia*, is so called because of the ear shape of the large opening to the shell. The shell itself is rather squat and about 25 mm high. This species is found in large ponds, lakes and slow-moving waters. The mud snail†, *L. glabra*, has a quite characteristic, almost cylindrical shell,

 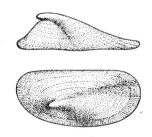 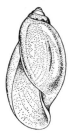

Fig. 51 Pulmonates: *Left: Lymnaea peregra*, 12–18 mm high. *Middle: Ancylus lacustris*, about 4 mm long. *Right: Physa fontinalis*, 6–7½ mm broad

12–25 mm high. The species is often found in shallow grassy ditches and temporary ponds which dry up in hot weather.

The freshwater limpets, of which there are two species in Britain, have quite a different type of shell from those described above. It is hood-shaped with no whorls, but the blunt apex is tilted. The limpets are usually found on stones or plants, and so firmly do they attach themselves when alarmed that it is often difficult to remove them without damaging the shell. When they move, very little of the body is protruded and the shell is swung slowly from side to side. Although pulmonates, they do not need to rise to the surface for air, but seem able to extract all they need from the water.

The river limpet†, *Ancylastrum fluviatile* (= *Ancylus fluviatilis*), is very common in most parts of the country, attached to stones in streams, clear lakes and tarns. Its shell is oval and 6–8 mm long, with the apex twisted to the right.

The lake limpet*, *Ancylus lacustris* (= *Acroloxus lacustris*), which has a more restricted distribution, has a flatter, smaller and narrower shell, and the apex is turned to the left. This species frequents stiller waters, although in some districts the two may be found together. It is usually found on water-plants. The eggs of both species are deposited on stones or other submerged objects and enclosed in a gelatinous covering. (Plate **24, 7**)

The bladder snail†, *Physa fontinalis*, is a common species in clear ditches and streams. Its thin oval shell is about 12 mm high. (Plate **24, 11**) The edge of the mantle has lobes which can be wrapped round the shell to serve as gills.

Most people are familiar with the ram's-horn or trumpet snails, and they are favourite molluscs to keep in aquaria. The fourteen species range in size from the great ram's-horn* (Plates **22, 3; 24, 10**), over 25 mm in breadth, to the nautilus ram's-horn, *Planorbis crista*, only 3 mm in breadth; in all, the shell is coiled in a flat spiral. The foot is small and rounded at each end, and a pair of long, thin, tapering tentacles with eyes at their base are carried on the head.

Unusual in molluscs, the blood of ram's-horn snails contains haemoglobin and is therefore red in colour. The combination of the haemoglobin, a large breathing aperture and what appears to be a simple gill in the form of a small lobe of skin on the snail's left side, make the great ram's-horn better adapted to an aquatic existence than, for example, the great pond snail, and in well-oxygenated waters it rarely needs to surface for air. The red specimens sometimes found and so highly prized by aquarists are albinos which lack the normal brown pigment in the shell and body, so that the red blood shows through and makes the snail a vivid crimson colour.

The eggs are enclosed in a roughly circular gelatinous capsule on stones or wrapped round plant stems. When deposited on the flat sides of an aquarium they take the shape seen in Plate **23, 2**.

The great ram's-horn*, *Planorbarius corneus*, is the best-known species, and its shell, which may measure over 25 mm in breadth, has five or six convex whorls. It is somewhat local in its distribution, and is found in rivers, canals and lakes. The ram's-horn*, *Planorbis planorbis*, is common in all parts of the country in ponds, ditches and slow-moving waters. Its shell is smaller than the great ram's-horn, being about 18 mm in breadth. The remaining common species are smaller, and are ususally found among water-plants. (Plate **24,** 3)

The Freshwater Mussels and Cockles

Class *BIVALVIA*

The bivalves or lamellibranchs are represented in fresh water by the larger mussels, the smaller orb-shell and pea-shell cockles and the somewhat alien zebra mussel, the total comprising twenty-six species.

The characteristic features of a bivalve can best be seen in one of the larger species, such as the swan mussel, *Anodonta cygnea* (Fig. 52). The shell consists, as in all freshwater species, of two equal valves, hinged at the top by an elastic ligament. Powerful muscles are present inside to close the shell tightly when required, and other muscles attach the body to the shell. The oldest part of the shell is the humped area in front of the hinge, called the *umbo*. In some genera

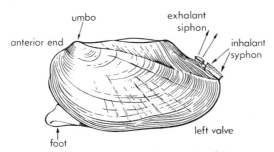

Fig. 52   Structure of a bivalve, *Anodonta*

the two valves are held together with hinge-teeth that fit into grooves or pits in the opposite shell, but these are not present in *Anodonta*.

Bivalves have no head, and only the foot is protruded through the front part of the shell. In the ordinary way the front two-thirds of a swan mussel is embedded in the mud at the bottom of the water, and the valves are held slightly ajar. The foot is sunk into the mud at the front end of the shell, and at the rear are the openings of two tubes which can be well seen in a specimen kept in an aquarium. The uppermost and smaller of these is called the *exhalant siphon,* and the lower one, fringed with finger-like outgrowths, the *inhalant siphon.* (Plate **23,** 4) By means of these a continuous current of water is drawn through the mantle-cavity of the mussel and expelled, bringing not only supplies of oxygen for breathing, but also the floating particles of algal food and minute animals on which the creature lives. The current is created by four spongy gills which lie in pairs along the middle of the body on either side of the foot. The gills bear cilia on their complex network, and it is the lashing of the cilia that causes the current. The food particles in the current are trapped in a sticky secretion of the gills as the water passes through them, and are carried by ciliary action in a slimy mass to the mouth of the animal at the base of the front part of the foot. Respiration is not carried out exclusively by the gills as was formerly thought, for it is now known that the mantle also takes part in this process.

Mussels seem able to control what enters the body and, if some obviously undesirable substance is placed near, it will be noticed that the inhalant siphon *ejects* a stream of water to drive the offending matter away.

The smaller bivalves are hermaphrodite, but the larger ones have distinct sexes, although occasionally hermaphroditic individuals are found. The females produce many thousands of eggs during the summer months, and these are passed from the ovary into brood-pouches in the outer gills, where they are

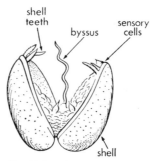

Fig. 53   Glochidium larva of *Anodonta* (× 100)

retained. Sperm from another mussel enters through the inhalant siphon and fertilizes the eggs, but they do not complete their development until the following spring, when the larvae, of a peculiar type called a *glochidium*, are shot out into the water. The *glochidia* have a miniature bivalve shell with a sharp tooth at the free apex of each triangular valve. The valves bear minute perforations, and from the muscle which enables them to open and close arises a coiled, sticky thread, the *byssus*. After they are ejected from the parent, the small glochidia (Plate **23, 3**) swim about for a time by snapping their tiny valves together, eventually falling to the bottom or becoming entangled by their byssal threads in water-plants. Their further development depends on becoming attached to a fish, for at this stage they become parasites. Although many thousands must fail to find a host, the fortunate ones manage to attach themselves to a fish which is perhaps swimming among the plants in which they are entangled. The byssa of many glochidia may stick to the fish and be carried away on its body. Those that can, then embed themselves in the skin of the fish by means of their sharp teeth, usually on the fins or tail, and this causes the tissues of the fish to form a cyst round each larva. In this condition the young mussels remain for about three months, living as true parasites and obtaining their food from the blood of the fish. A new shell develops under the old one, and finally the perfectly formed little mussels drop off their host and lead a free existence. It is a strange fact, however, that very few of these minute mussels are ever found.

Although many species of freshwater fish become hosts to mussel larvae, the three-spined stickleback seems particularly prone to attack, and sometimes a single stickleback will have many little swellings, each indicating the presence of a glochidium, on its fins or tail. It is interesting to remember that one fish at least, the bitterling, a common fish on the Continent and one popular with aquarists here, turns the tables on the mussels and uses one species in particular, the painter's mussel, *Unio pictorum*, as a nursery for its young. The female deposits her eggs into the inhalant siphon of the mussel by means of her long ovipositor, and the fertilizing milt of the male is also discharged near by, so that it, too, finds its way inside the mussel. The fertilized eggs develop there and the young fish remain until they are ready to lead an independent life, when they emerge into the water.

The larger mussels are sometimes kept in aquaria, and serve as very efficient filters, removing the floating algae which tend to make the water cloudy. In view of their unfortunate habits of ploughing a way across the tank, irrespective of the choicest water-plants in their path, and of dying unobtrusively and polluting the water, many aquarists consider they are more bother than they are worth!

The Larger Mussels

All the larger freshwater mussels belong to the super-family *Unionacea*, and although the shell is usually sufficiently characteristic to make possible the identification of each genus, considerable variations in colour, shape and size do occur, particularly in specimens from different localities, and features such as the presence or absence of hinge-teeth and their type, the form of siphon tubes, and the structure of the gills (which, of course, can only be seen in a dissection) are relied on for more detailed naming.

The pearl mussel†, *Margaritifer margaritifer* (= *Unio margaritifer*), is found in swift rivers in the north of England, along the west coast, and in Scotland and Ireland. The shape of the shell varies a great deal, but is usually long and oval and of a blackish colour. It is from 80–160 mm long and bears hinge-teeth.

It is perhaps not generally known that there are any pearl-producing molluscs in Britain. Small pearls of little value are sometimes found in other members of the *Unionacea*, but those of the pearl mussel are sometimes quite large, and British pearl fisheries were famous in Roman times. More recently, important fisheries were run profitably, particularly on the River Tay in Perthshire, from where it is stated that £10,000 worth of pearls were sent to London between the years 1761 and 1764. The pearls of freshwater molluscs, which may be white, pink, green or brown, are formed in the same way as those of the marine oysters, i.e. by the outpouring of nacre from the mantle to surround a foreign body such as an encysted larva of a trematode fluke, the adult of which is said to live in swans and tufted duck.

The remaining four mussels live in slow rivers, canals, lakes and large ponds. The painter's mussel*, *Unio pictorum*, has a long thin shell, from 80–100 mm long with straight upper and lower margins, the hinge and hinge-teeth being slender. The body of the animal is reddish in colour. Both the common and scientific name (L. *pictor*, a painter) were given to the species because the early Dutch painters used the valves of the shell to hold their colours. Until recently gold and silver leaf used for illuminated manuscripts was sold in these shells.

*Unio tumidus** is not so abundant as the previous species, and frequents cleaner water. The shell is swollen and the margin is rounder than in *U. pictorum*. Both species of *Unio* are confined to England and Wales.

The swan mussel*, *Anodonta cygnea*, has already been described. Its oval, greenish-yellow shell may reach a great size, and specimens up to 230 mm long have been recorded, but the average is about 136 mm. There are no hinge-teeth. The duck mussel*, *Anodonta anatina*, is frequently found with *A. cygnea*, and it is not always easy to distinguish between the two species, for they are both variable in their characteristics. On the whole, the shell of the duck mussel is smaller, thicker and darker in colour, and the umbo is placed more centrally. (Plate 25, 4, 5)

Orb-shell Cockles

The four species of orb-shell cockles belong to the genus *Sphaerium*. All are small molluscs with pale-coloured shells which can easily be mistaken for small pebbles. The valves are rounded, with the umbo placed almost in the middle of the hinged margin, so that the whole shell is almost spherical. The shell bears hinge-teeth. Orb-shell cockles are hermaphrodite, and the young, when ejected from the shell of the parent, are fully developed and do not go through a parasitic stage.

*Sphaerium rivicola** is the largest species, with a shell which may be up to 27 mm in length. It inhabits slowly flowing rivers and canals, and is widely distributed in England and Wales.

S. corneum is the commonest orb-shell, and is found all over the British Isles in most kinds of freshwater habitat. Although usually found on the bottom mud or gravel, it is sometimes taken among water-plants, for it frequently climbs up the stems. The shell varies greatly in size, depending on the particular habitat, but is usually about 11 mm long. *S. transversum* is found in canals and slow rivers, while *S. lacustre* inhabits stagnant water. Both are about 12 mm in length. (Plate 25, 1)

Pea-shell Cockles

The fifteen British species of pea-shell cockles all belong to the genus *Pisidium*. Their shells are smaller than the orb-shell cockles and not so symmetrical, since the umbo is not in the middle of the hinged margin, but towards the rear. A further distinguishing feature seen only in specimens that are active is that, whereas *Sphaerium* has two siphon tubes protruding, *Pisidium* has only one. The habits of the members of this genus are similar to those of the orb-shell cockles, and like them are favourite foods of fish, which swallow the molluscs, shell and all. Ducks and other water-birds, too, are very fond of them. (Plate 25, 2)

Frequently, the feet of water-birds have attached to them specimens of both kinds of cockles, their valves tightly closed and gripping the skin. In this way the molluscs are no doubt distributed widely, not only within this country, but to and from other countries. The mode of reproduction is the same as in the orb-shell cockles.

It is not easy to distinguish between the various species of *Pisidium*, and only a few of the best-known species will be mentioned briefly.

Pisidium cinereum† is probably the commonest species, and perhaps the most widely distributed mollusc in this country, inhabiting all kinds of freshwater habitats. The shell is a little over 6 mm in length.

P. milium† is almost as ubiquitous as the last species, and is found in most types of water except swamps. *P. amnicum*, the largest species, may be 10 mm in length. It is found in clean running waters and lakes, and is generally distributed in England but infrequently in Scotland. *P. personatum*†, about 4 mm in length, also inhabits shallow ditches, temporary pools and swampy places.

The Zebra Mussel

To complete our survey of the freshwater bivalves mention must be made of the zebra mussel, *Dreissena* (= *Dreissensia*) *polymorpha*, which was first noticed in this country in 1824, and is believed to have been brought in timber ships from the Baltic. It is more like the marine mussels than the freshwater species, and has their habit of attaching itself in groups to submerged objects by means of sticky byssal threads.

The shell, which is without teeth, is yellow or brownish, handsomely marked with wavy or zigzag transverse brown or yellow bands. The length of the shell may be 50 mm, but much smaller specimens are often found. (Plate 25, 3)

The zebra mussel is widely distributed in England and south Scotland, and is usually found in slow rivers, canals, docks or reservoirs, attached in clusters to submerged stones or wooden posts. It has now become a rather troublesome pest of the water-engineer by blocking water pipes, but it has been suggested recently that the ability of mussels to extract bacteria and other micro-organisms and floating organic matter, as well as taking up and concentrating heavy metals polluting the water, might be used by setting up beds of *Dreissena*, and other selected shellfish, as a first stage of purification for many industrial and domestic effluents.

Dreissena shows its affinities to the marine bivalves by having a larval stage, called a *veliger*, as they have. When first hatched from the egg, it bears a crown of cilia, and for a time swims freely in the water by means of these. It does not pass through a parasitic stage, but after a few days of this free-swimming existence, sinks to the bottom, grows a shell and attaches itself to a solid support with its *byssus*.

Although the zebra mussel has increased rapidly wherever it has gained a footing, it is doubtful if it could ever establish itself in fast-flowing water, as the free-swimming larvae would be swept away.

DRAGONFLIES

1 Brown aeshna, *Aeshna grandis*. 2 Golden-ringed dragonfly, *Cordulegaster boltonii*. 3 Common aeshna, *Aeshna juncea*. 4 Broad-bodied libellula, *Libellula depressa*. 5 Emperor dragonfly, *Anax imperator*. 6 Four-spotted libellula, *Libellula quadrimaculata*. 7 Southern aeshna, *Aeshna cyanea*. 8 Common sympetrum, *Sympetrum striolatum*. (Approximately half natural size)

Pl. 33

Pl. 34

MAYFLIES

1 Flattened type of nymph from a stream, *Ecdyonurus* (× 3). 2 Egg clusters of *Baetis* underneath a submerged stone (× 3). 3 Sub-imago of *Ephemera* (× 1½). 4 Tracheal gills of *Cloëon* (× 3). 5 Burrowing type of mayfly nymph, *Ephemera danica* (× 4). 6 Imago just emerged from the sub-imaginal skin (× 1)

Pl. 35
STONEFLIES

1 Nymph of *Perla bipunctata* (× 4). 2 Cast nymphal skin of a stonefly on the trunk of a tree. 3 Adult stonefly and the nymphal skin from which it emerged (× 2)

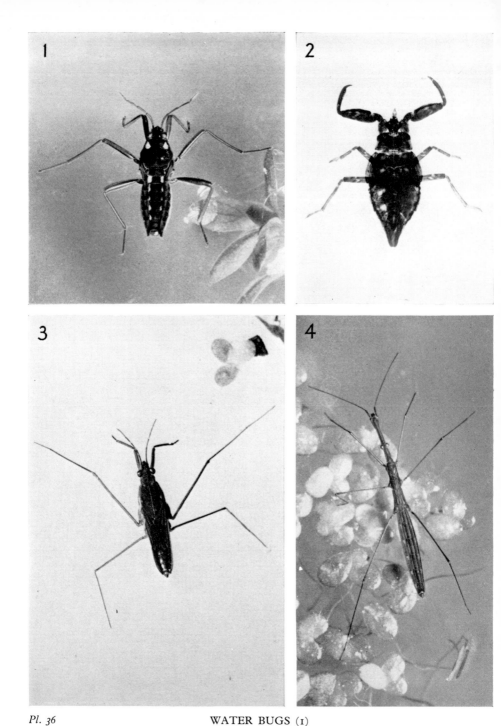

Pl. 36

WATER BUGS (1)

1 Water cricket, *Velia caprai* (× 5). 2 Nymph of water scorpion, *Nepa cinerea* (× 5). The long breathing-tube has not yet developed. 3 Pond skater, *Gerris najas* (× 2). 4 Water measurer, *Hydrometra stagnorum*, on the surface among duckweed, *Lemna minor* (× 5)

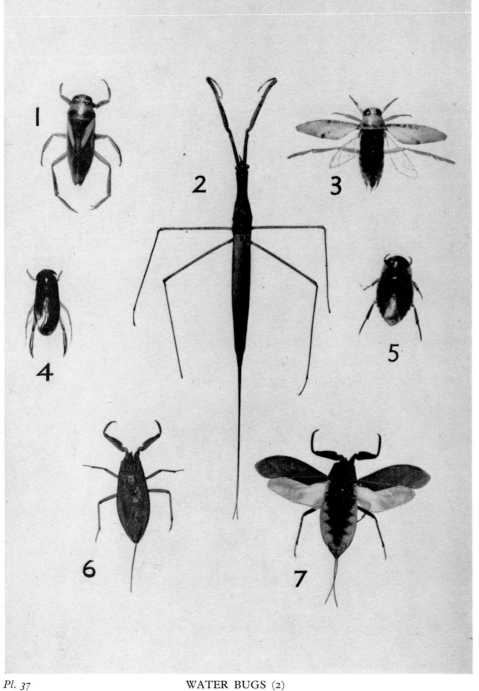

Pl. 37   WATER BUGS (2)

1 and 3 Water boatman, *Notonecta*. 2 Water stick insect, *Ranatra linearis*. 4 Lesser water boatman, *Corixa*. 5 Saucer-bug, *Ilyocoris* (= *Naucoris*) *cimicoides*. 6 and 7 Water scorpion, *Nepa cinerea*. (All × $1\frac{1}{4}$)

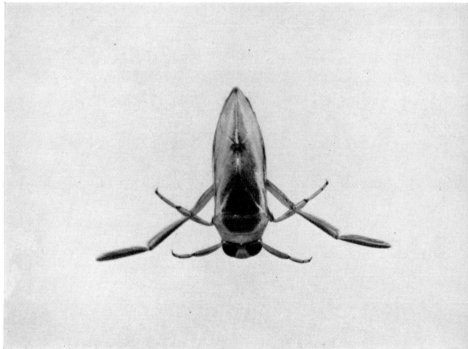

Pl. 38   WATER BUGS (3)

Above: Aphelocheirus montandoni (× 5). *Below:* Water boatman in swimming position
with the long, fringed swimming legs well shown

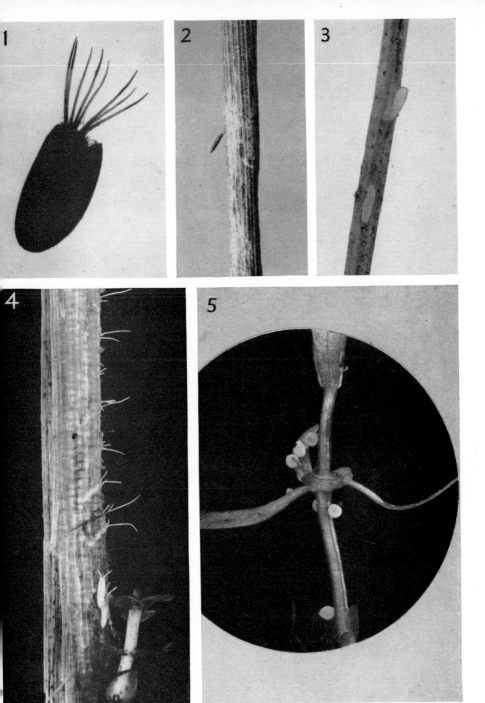

Pl. 39 EGGS OF WATER BUGS

1 Water scorpion, *Nepa cinerea* (× 20). 2 Water measurer, *Hydrometra stagnorum* (× 6). 3 Water boatman, *Notonecta* sp. (× 12). 4 Water stick insect, *Ranatra linearis:* two whole eggs are visible below; the pairs of filaments of others may be seen along the stem (× 5). 5 *Corixa* on starwort, *Callitriche* (× 4)

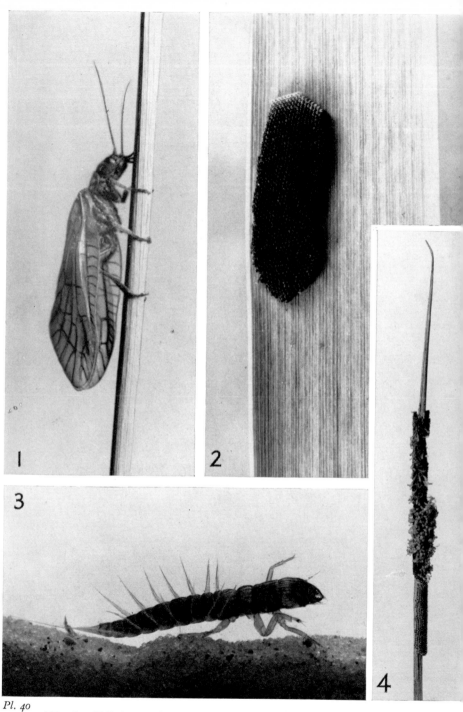

Pl. 40

1 Alder-fly, *Sialis lutaria*, on reed stem (× 4). 2 Egg cluster of alder-fly on the stem of a bur-reed, *Sparganium* (× 6). 3 Larva of *Sialis lutaria* under water (× 5). 4 Egg cluster of *Sialis* with larvae hatching (× 3)

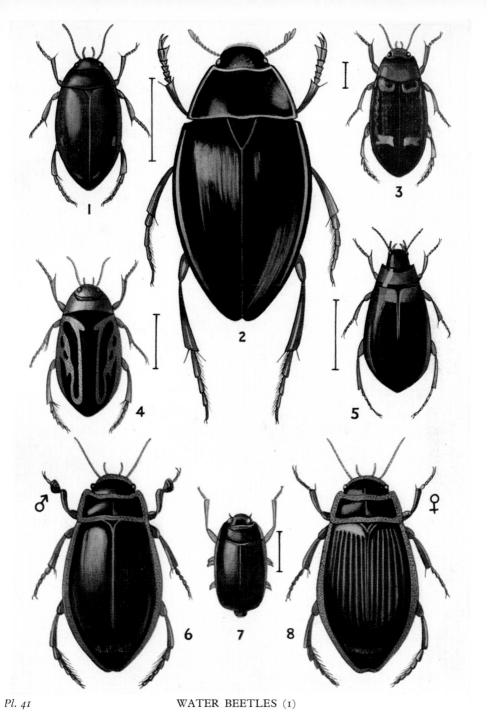

Pl. 41   WATER BEETLES (1)

1 *Ilybius ater*. 2 Great silver beetle, *Hydrophilus piceus*. 3 *Laccophilus variegatus*.
4 *Platambus maculatus*. 5 Screech beetle, *Hygrobia hermanni*. 6 and 8 Great diving
beetle, *Dytiscus marginalis*. 7 Whirligig beetle, *Gyrinus natator*. (Nos 2, 6 and 8, and all
scale lines twice natural size)

1 Great diving beetle, *Dytiscus marginalis*: *left*, male; *right*, female (× 1½). 2 Eggs of *Dytiscus* embedded in stem of *Poa aquatica* (× 5). 3 Fore-leg of male *Dytiscus marginalis* showing the sucker-pad (× 35). 4 Larva of *Dytiscus marginalis* under water (× 1)

Pl. 42

Pl. 43   1 Egg-case of great silver beetle, *Hydrophilus piceus* (× 1½). 2 *Hydrophilus piceus* at the surface, showing the silver air bubble (× 1). 3 Larva of whirligig beetle, *Gyrinus* sp. (× 7). 4 Larva of beetle *Acilius sulcatus* at the surface of the water (× 3). 5 Larva of *Hydrophilus piceus* under water (× 2)

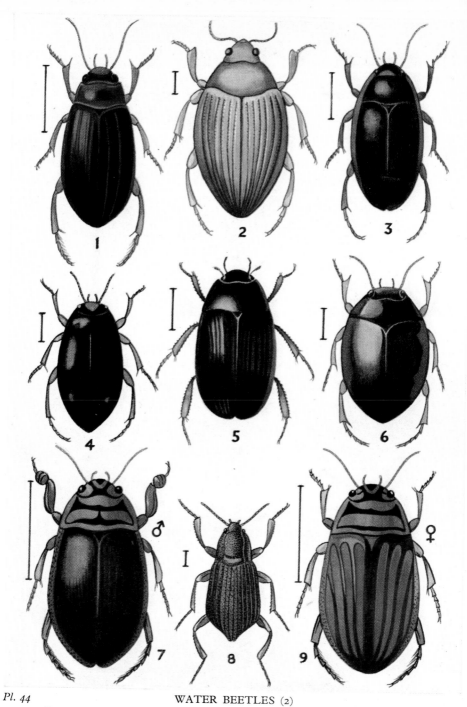

Pl. 44 WATER BEETLES (2)

1 *Rantus notatus.* 2 *Haliplus flavicollis.* 3 *Agabus uliginosus.* 4 *Hydroporus erythro-cephalus.* 5 *Hydrobius fuscipes.* 6 *Hyphydrus ovatus.* 7 *Acilius sulcatus*, male. 8 *Elmis aenea.* 9 *Acilius sulcatus*, female. (Scale lines twice natural size)

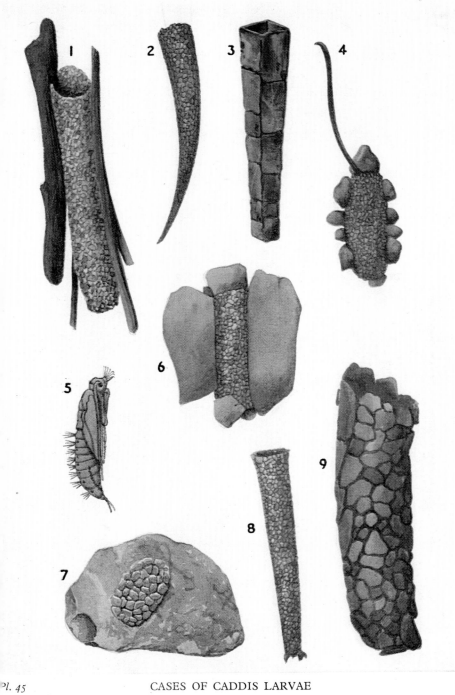

Pl. 45

CASES OF CADDIS LARVAE

1 *Anabolia nervosa*. 2 *Athripsodes* (= *Leptocerus*) *aterrimus*. 3 *Lepidostoma hirtum*. 4 *Silo pallipes* (parasitized by the ichneumon fly, *Agriotypus armatus*). 5 Pupa of *Goera pilosa*. 6 *Goera pilosa*. 7 *Agapetus fuscipes* attached to a stone. 8 *Sericostoma personatum*. 9 *Potamophylax latipennis*

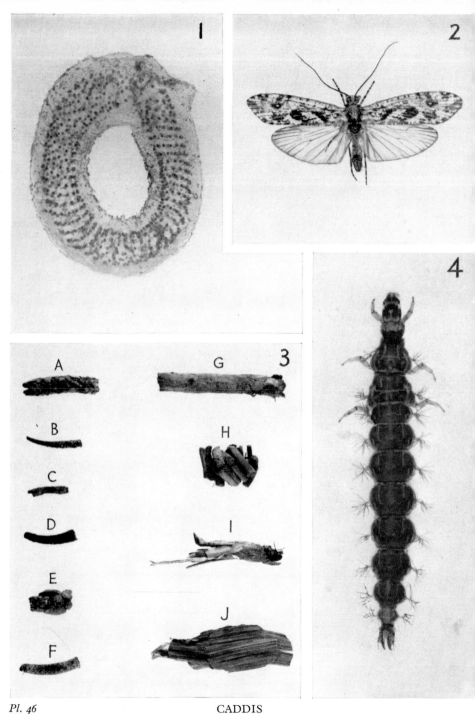

Pl. 46

CADDIS

1 Egg-rope of *Phryganea grandis* (× 3). 2 Adult caddis-fly, *Phryganea striata* (× 3).
3 Caddis cases: A. *Potamophylax*; B. *Leptocerus*; C. *Lepidostoma*; D. *Odontocerum*;
E. *Silo*; F. *Sericostoma*; G. *Phryganea grandis*; H. *Limnephilus*; I. *Anabolia*;
J. *Glyphotaelius*. 4 Non-case-building larva *Rhyacophila*; note the tracheal gills along
the sides of the body (× 5)

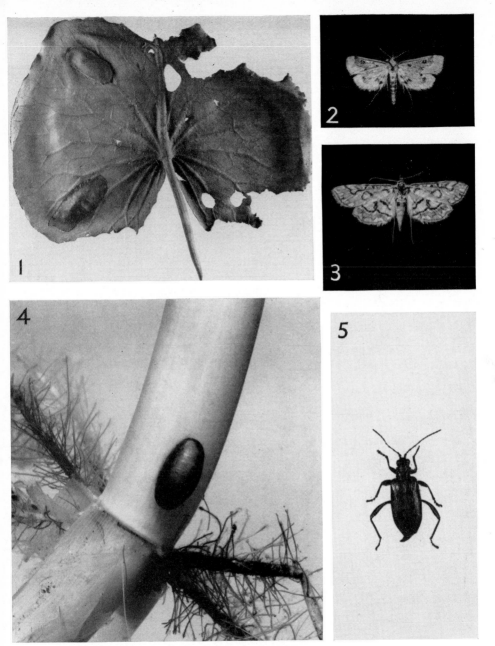

Pl. 47
  1 Two cases made by caterpillars of the china mark moth *Nymphula nymphaeta* attached to a water-lily ($\times \frac{1}{2}$). 2 and 3 China mark moths; *above: Cataclysta lemnata*; *below: Nymphula nymphaeta* ($\times$ 2). 4 Pupal case of the beetle *Donacia* on the lower part of a submerged reed stem ($\times$ 2). 5 Adult beetle, *Donacia crassipes* ($\times$ 2)

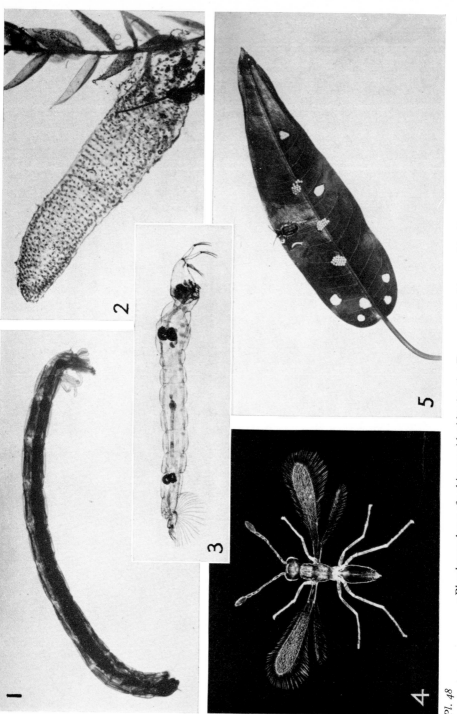

1 Bloodworm larva of a chironomid midge ($\times$ 7). 2 Egg-rope of a chironomid midge ($\times$ 7). 3 Phantom larva, *Chaoborus crystallinus* ($\times$ 10). 4 Chalcid wasp, *Caraphractus cinctus* ($=$ *Polynema natans*) ($\times$ 15). 5 Adult and three egg-clusters of beetle *Galerucella grisescens* on the underside of a leaf of amphibious persicaria, *Persicaria amphibium* ($\times$ 1½)

Pl. 48

1 2: THE TRUE WORMS

Perhaps the most familiar example of the phylum Annelida (Fr. *annelide* <
anneler, to arrange in rings) is the familiar earthworm of our gardens. If the
long, soft, cylindrical body is closely examined, it will be noticed that it is marked
off all along its length by equally spaced rings. In the earthworm and its close
relatives, the rings correspond to separate compartments inside the body, each
containing an almost complete and similar set of internal organs.

The annelids show a distinct advance in having a true body-cavity, or *coelom*,
between the food canal and the body-wall, although in the leeches this space is
filled up with muscles and connective tissue. Some have blood containing the
red substance haemoglobin, which is present in our own blood.

The skin of an annelid is covered with a soft cuticle, not chitinized like that
of insects and crustaceans, and many of them possess bristles, or setae, arranged
in groups on the segments of the body, which assist them in moving about or in
gripping the sides of their burrows. Anyone who has watched a thrush trying to
pull a worm out of the lawn will realize how effective the bristles are for this
purpose.

Annelids are divided into two major groups: *Chaetopoda* (Gr. *chaite*, a bristle;
podos, a foot), which includes the worms with bristles, and *Hirudinea* (L. *hirudo*,
a leech), the leeches, which have no bristles.

The *Chaetopoda* are again divided into the classes *Polychaeta* (Gr. = many
bristles) and *Oligochaeta* (= few bristles). The former are all marine worms and
the freshwater annelids belong to the latter class.

Class *OLIGOCHAETA*

The freshwater oligochaetes are very common in both still and running waters.
Often they live in the mud at the bottom, where some make burrows, and the
best way to obtain specimens of these for study is to scoop up quantities of mud
and spread it out in a thin layer over the bottom of a shallow dish.

Other species will be found by pulling apart the decaying stems of such
plants as reed-mace and bur-reed, and still others occur in masses of algae. Many
kinds are transparent and almost colourless, so that a careful examination is
sometimes necessary to see them, even in the restricted space of a dish. The
transparent species are particularly interesting creatures to study under a low

K

power of the microscope, for their internal organs can be seen with ease. A live-box (see p. 244) will be needed to hold the specimens in position, for they are lively animals and would otherwise soon wriggle out of the field of view.

Although each individual oligochaete has both male and female reproductive organs, cross fertilization between two worms is necessary to produce fertile eggs. These are enclosed in a capsule or 'cocoon', formed by a structure called the clitellum which appears as a swollen area near the front of the body of the worm. At first the cocoon encircles the body, but after passing over the front end

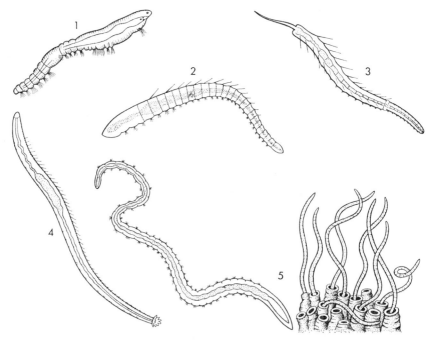

Fig. 54 Annelid worms: 1 *Chaetogaster*, 8 mm. 2 *Nais*, 3–25 mm. 3 *Stylaria*, up to 18 mm. 4 *Dero*, 10–15 mm. 5 *Tubifex*, up to about 8 cm; *right:* part of a colony in the mud

of the creature the ends close, to form capsules that differ in shape in the various species and are either deposited on plants or dropped to the bottom of the water. There is no larval stage as there is in polychaete worms, and the eggs give rise to fully-formed young.

Asexual reproduction by lengthwise budding also commonly takes place in the freshwater oligochaetes, and the process may be repeated several times before the newly-formed worms separate, so that a chain of individuals is often seen.

The food of aquatic worms consists chiefly of decaying vegetable matter in the mud and, like their earthworm relatives, they pass vast quantities of material through their bodies to obtain what they require, thus breaking down the organic matter into a finely divided state and aiding in its disintegration.

There are nine families of freshwater oligochaetes which are fairly easily recognizable on sight; the form and arrangement of the bristles are the main features on which more detailed classification is based, although examination under a high power of a microscope may be needed to resolve the finer points. The best-known families are:

LUMBRICIDAE   This is the family to which belong our common earthworms. An aquatic species, the square-tailed worm, *Eiseniella tetrahedra*, resembles a small earthworm, and is about 50 mm long with the rear end square in section. It is found in the mud at the margin of lakes and also among vegetation near the water's edge of streams.

NAIDIDAE   The members of this large family are, perhaps, the commonest of all the freshwater annelids. They are all small creatures, rarely exceeding 25 mm in length, some species being only about 3 mm. Often chains of several individuals may be joined for, in this family, multiplication by budding takes place more frequently than by the laying of eggs.

Chaetogaster limnaei is a colourless worm, about 8 mm long, sometimes found inhabiting the occupied shells of snails or the tubes of 'bloodworms', *Chironomus*. *C. diaphanus*, which may be up to 25 mm long, is very transparent, and lives freely among water-plants. The genus *Nais* contains nine species which often appear among collections of weed and algae, or among detritus at the bottom of the water. Single worms are up to about 25 mm long, but it is common to find several joined in a chain. The bristles on the back are particularly long in this genus.

Stylaria lacustris, which also has long bristles on the back, can be distinguished by its long, thin proboscis. The species in the genus *Dero* are slender worms about 18 mm long and pinkish in colour; they live in mud tubes. At the tail end are leaf-like gills which can be retracted or expanded at will. Members of the genus *Ophidonais* sometimes reach a length of 30 mm, and are the largest representatives of the family. The head end of their bodies bears broad stripes.

TUBIFICIDAE   On the mud in the shallow area at the margins of ponds and slow streams a reddish smudge will often be seen. The vibration made by an incautious step will cause it to disappear, and all that will be left for a few moments will be tiny pits in the mud surface. Soon, if we keep still, the 'tails' of innumerable red worms will reappear, and by their waving to and fro restore the red patch to view. These creatures are river-worms, the commonest genus being *Tubifex*. They live head down in tubes made in the mud, waving the posterior region of the body ceaselessly to obtain oxygen from the water. *Tubifex* worms are well adapted for living in waters deficient in oxygen. Not only does the waving motion create currents, which enables the worms to abstract this gas from a comparatively wide expanse of water, but the blood of the creatures contains haemoglobin, the red, respiratory pigment which is in our own blood. The great affinity of this substance for oxygen ensures that the most is made of the limited supply often available in the conditions under which these worms live. If the oxygen supply in the water becomes less, the worms come farther out of their burrows to increase the absorbing area of the body.

Tubifex worms lay their eggs in oval capsules and breed very rapidly. Large numbers of these creatures are collected and bred by aquarists for feeding their fish.

LUMBRICULIDAE   *Lumbriculus* is a very common worm in still or slow-moving waters. Specimens up to 75 mm or more are found, the general colour being red or brown, but a greenish pigment obscures this colour in parts of the body. *Lumbriculus* multiplies by division, a complete worm developing from each portion.

ENCHYTRAEIDAE   The members of this large family are often called 'pot-worms'. About 12 mm long, they are usually found among the roots of water-plants, and on account of their white colour can easily be mistaken for the rootlets themselves. The species of *Enchytraeus* are very common, some being as much at home in damp conditions on land as in the water. *Lumbricillus* (not to be confused with *Lumbriculus*) is larger, often 25 mm or more in length, and of a somewhat similar appearance to an earthworm.

The freshwater oligochaetes perform useful functions with their scavenging activities and in providing an abundant source of food for the larger animals including fish.

The Leeches: Class *HIRUDINEA*

It is, perhaps, inevitable that the blood-sucking activities of some kinds of leech and their ability to attach themselves firmly to other animals by means of suckers, have not endeared them to the human race, but there is little justification for this evil reputation in the British species, only one of which, and that a rare one, being capable of penetrating the human skin.

Leeches are very common creatures in all types of fresh water, and although at first sight they do not bear much resemblance to the other annelids, they share with them many features of external and internal structure. On the other hand, they have no bristles, their bodies are flattened and more oval in shape when at rest, and the rings that mark the surface do not correspond with the internal segments, which are fewer. The body-cavity, too, is filled with muscle and connective tissue, so that the body would appear solid if it were sliced through.

The most characteristic features of the leeches are the two suckers, a small one at the head end surrounding the mouth and not always easy to see, and a large one at the rear end. The large sucker enables the creatures to hold on to surfaces and to its prey, the smaller one serving mainly to obtain a firm grip of the victim while feeding. Most leeches are blood-suckers, attacking fish, frogs, snails and insect larvae. Some species pierce the skin of their victim by means of a proboscis, but others have jaws and inflict a Y-shaped wound. Secretions from the leeech's salivary glands prevent the coagulation of the victim's blood and ensure that it flows freely and does not clot when it is stored in the leech's body. Once having obtained a hold on its prey, a leech is almost impossible to detach, and not until gorged with blood does it drop off. A medicinal leech can take three times its

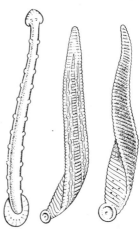

Fig. 55 Leeches : *Left: Piscicola geometra*, 25–50 mm. *Middle: Hirudo medicinalis*, 10–15 cm. *Right: Haemopis sanguisuga*, 10–15 cm

own weight of blood at a single meal. The digestive tract has a series of pouches in which the blood is stored, and one meal may last some months.

Leeches are not active creatures and will usually be found attached to stones or among water-plants, their powerful suckers enabling them to live in even fast-flowing waters. In the ordinary way they move about by using their suckers. First of all, the body is stretched out, often to a great length, and waved around until a suitable point of attachment is found for the front sucker. The rear sucker is then detached and the body contracted to enable it to secure a hold near the front one, and so the process continues. Some leeches can swim easily with graceful undulations of the body, and often do so if disturbed.

All leeches have both male and female organs, and although each individual is therefore capable of laying eggs, self-fertilization has not been observed. In most species, fertilization is effected by the injection of the spermatophore, a lance-shaped capsule containing sperms, into the skin of another individual. The sperms pass into the tissue in the body-cavity and thus to the ovaries to fertilize the eggs. Several spermatophores are sometimes found attached to the body of a single leech, and they adhere to the skin for some time after fertilization has taken place.

After fertilization has taken place the eggs of most leeches are deposited in cocoons, which are attached to submerged stones or other objects, although some species deposit them out of the water. During the breeding season, April to September depending on the species, the cocoon-depositing species develop a short swelling—the clitellum—near the front part of the body. Skin-cells in this region secrete a delicate cylindrical envelope which passes by muscular action towards the head. When it reaches the genital pore, the fertilized eggs are enclosed in it, and it then passes over the head end and so off the body. The ends close up and the cocoon becomes oval in shape, and in this form is firmly cemented to a suitable support.

In contrast to this method of egg-deposition, members of the family Glossiphoniidae attach their eggs singly to the under-surface of their own bodies, or deposit them in delicate capsules on to solid supports and there brood over them. The young, when hatched, attach themselves to the under-surface of their parent, and as many as 215 young have been observed at once on the body of an adult *Theromyzon tessulatum*.

Fourteen species of freshwater leeches are found in Britain, varying in length, when at rest, from 10 mm to as much as 100 mm. When extended, they may be two or three times these sizes. The colours vary a great deal, not only between the species but even in the same individual, for, by means of pigment cells in the skin, leeches can to a limited extent change their colour to suit the conditions in which they find themselves. Identification is based mainly on the number and arrangement of the eyes and the shape of the body, but the general behaviour of the creatures when moving is often a help to those familiar with them.

Leeches with a Proboscis and no jaws

Sub-order RHYNCHOBDELLAE (Gr. *rhynchos*, a beak, snout; *bdella*, a leech)

PISCICOLIDAE (L. *piscis*, a fish; *colo*, to inhabit) The species included in this family prey on freshwater fish and usually attach themselves to the fins and cloacal region. They may remain on their host for two or three days, drawing its blood, and dropping off only when gorged. *Piscicola geometra* is fairly common in both still and running water. It has two pairs of eyes, and its attenuated, semi-transparent body is about 25 mm long, brownish-green in colour with rows of brownish spots around the body. Both suckers are large and distinct from the body. The eggs are laid in dark-brown oval cocoons attached to submerged stones and plants.

GLOSSIPHONIIDAE (Gr. *glossa*, tongue; *siphon*, tube) The members of this family are small leeches, some of which prey largely on water-snails, and as a consequence are sometimes referred to as 'snail-leeches'; but they feed also on worms and insect larvae, such as bloodworms. *Theromyzon tessulatum* is found attached to the throat and nasal cavities of water-fowl. Although uncommon, it has a wide distribution, no doubt due to this habit of attaching itself to birds. Greenish-grey in colour, with six rows of yellowish spots on the back, it is about 25 mm long when at rest, and has four pairs of eyes. The eggs are deposited in capsules attached to submerged objects, or sometimes carried about on the underside of the parent. A very common species of this family is *Helobdella stagnalis*, a small yellow- or greyish-coloured leech, only 10 mm long when at rest. Behind the single pair of eyes is a characteristic round, horny plate which serves to identify this species. Its eggs are carried about on the underside of the body. It feeds by sucking the body fluids of bloodworms and other aquatic invertebrates.

Glossiphonia heteroclita, and its larger relative *G. complanata*, are rather sluggish leeches, the former being yellow in colour and about 10 mm long, the latter nearly twice the size, and brown or green in colour, with two dark stripes along the back. Both have three pairs of eyes. *G. complanata* when disturbed usually contracts into a lump and drops off its support. They both feed by sucking the body fluids of water snails and are more commonly found in running water.

Leeches with Jaws
Sub-order GNATHOBDELLAE (Gr. *gnathos*, jaw)

HIRUDIDAE (L. *hirudo*, a leech)   In this family are included the largest and best-known leeches. The horse leech, *Haemopis sanguisuga*, is common all over the country, usually being found in the mud or under stones at the edges of ponds, ditches and sluggish streams. It occasionally leaves the water, and does so to deposit its egg-cocoons in the damp earth near by. About 60 mm long when at rest, it can extend to a length of 15 cm. The colour is very variable, dark green, light green and brown specimens occurring; both upper and lower surfaces bear black spots. There are five pairs of eyes.

The horse leech feeds mainly on snails, worms and tadpoles, but is not averse to small frogs and fish, or even members of its own species. It is not a blood-sucker and its prey is swallowed whole. In spite of its common name and the suspicion with which it is viewed by country folk, the creature is quite harmless to horses, for the teeth in its jaws are incapable of piercing the skin of human beings, let alone that of horses or cattle. There was an early superstition that nine horse leeches could kill a horse. (Fig. 55)

It is probable that in the past the horse leech was frequently confused with the medicinal leech, *Hirudo medicinalis*, which was once abundant in this country. From early times this species has been used for the 'bleeding' or 'blood-letting' of human beings, and it is the only British species capable of piercing the human skin. Nowadays this interesting creature is found only in a few localities including the New Forest, the Lake District, Anglesey, South Wales, Norfolk and Islay (Scotland). Leech-gathering was a profitable business in the early part of the nineteenth century, when venesection by leeches was a fashionable treatment for a variety of ills. When the native supply began to dwindle, large numbers had to be imported from the Continent, particularly from France; five million are said to have been received in England in 1824. Some idea of the scale on which leeches were used can be judged from the fact that nearly $57\frac{1}{2}$ million are stated to have been imported into France alone in the year 1832. No attempt at 'farming' the creatures appears to have been made in Britain, but on the Continent they were reared in large numbers. Even at the present time, leeches are said to be imported into the United States of America from Europe for use in the removal of black and blue spots, particularly 'black eyes'.

Although both adult and young leeches feed on frogs and small fish, it is believed that the adults cannot attain sexual maturity without access to mammalian blood. The closer control of the watering of cattle and horses with the coming of piped water supplies to all parts of the country, is probably mainly responsible for the rarity of the medicinal leech. It is significant that recent records of the animal have been from areas where stock is still allowed to roam.

The medicinal leech resembles the horse leech in general form and also in the number and arrangement of the eyes, but is usually olive-green in colour on the back surface with reddish-brown stripes and black spots. When at rest it contracts its body much more than does the horse leech. (Plate 17, 1)

Sub-order PHARYNGOBDELLAE (Gr. *pharynx*, the pharynx)

ERPOBDELLIDAE (Gr. *herpo*, to creep)   Two species of this family are found in this country: *Erpobdella octoculata* and *E. testacea*, 40 mm and 30 mm long respectively when at rest, brownish in colour and flattened. The latter species may be distinguished from the former by the absence of the characteristic black markings found on the back of *E. octoculata*. They both have four pairs of eyes.

Both leeches feed on worms, aquatic insect larvae and small crustaceans, swallowing their prey whole. *E. octoculata* is widely distributed in Britain in both running and still waters. *E. testacea* is rarer and usually found in reed swamps and organically polluted waters.

I3: THE CRUSTACEANS

The crustaceans, familiar examples of which are crabs, lobsters, crayfishes, shrimps and water-fleas, are included with the arachnids (spiders and mites), insects and myriapods in the phylum Arthropoda (Gr. *arthron*, a joint; *podos*, a foot), the largest group of animals, and the most successful in establishing itself over the face of the earth. This group colonizes almost every type of habitat in which life is possible, on land, in the sea and in fresh water.

Arthropods make up about three-quarters of the total animals living in fresh water, and in this chapter and the next three, the aquatic members of three of the classes mentioned above will be considered. Before dealing with the individual groups, a few remarks on arthropods in general will, however, be advisable.

The most characteristic features of a typical arthropod are the paired *jointed* appendages or limbs which are specialized and adapted for particular purposes. In early arthropods the appendages along the body were all alike and all served for several purposes, but in most present-day forms they are grouped so that one set is used for walking or swimming, another for grasping food and others for feelers or breathing organs. The more primitive present-day arthropods have bodies made up of more or less similar segments in much the same way as have the annelid worms, but in more highly developed forms the segments, although still present, differ from each other in structure and function.

The body of all arthropods is covered with a toughened layer, or cuticle, and can be considered as an external skeleton, serving as a framework for the internal tissues and a means of attachment for the muscles. It is composed largely of a horny substance called chitin (Gr. *chiton*, an outer covering), which is not made up of living matter but is secreted by the underlying skin cells. The outer surface of the chitin is generally covered with a thin layer of a waxy substance, which makes the cuticle waterproof.

Chitin itself is a flexible substance, and where it covers limb joints and spaces between the segments it allows free movement. Over other parts of the body, however, the chitin is often impregnated with lime salts, or other substances which make it hard and rigid. The hard, protective armour which surrounds arthropods has undoubtedly been one of the main reasons why these animals have become such dominant creatures in the earth today, but a non-living skeleton worn outside has one great disadvantage; it will not stretch to allow for growth, and thus all arthropods, to accommodate their growing bodies, have to moult the outer layer of chitin from time to time, a new one being secreted meanwhile underneath. During its development, an insect larva or nymph may have to moult twelve or more times before it becomes adult.

Most arthropods are active animals with a highly developed muscular system, a brain and nervous system and a body-cavity (*haemocoel*), containing blood. The blood of an arthropod does not circulate in a closed system of blood-vessels, but is pumped by the heart where present, into the head region, whence it finds its way through a series of spaces surrounding the main organs of the body and so back to the heart again. Not all arthropods, however, have hearts, e.g. cyclopoid and harpacticoid copepods.

Respiration is carried out in different ways in the various groups of aquatic arthropods. Some have gills to effect the interchange of gases with the surrounding water; others, being really land creatures that have returned to live in the water, take in atmospheric air through openings in their skin (*spiracles*), whence it is passed to the various parts of the body by means of air-tubes called *tracheae*.

After these preliminary remarks we can pass on to a consideration of the animals which form the subject of this chapter.

The members of the class Crustacea (L. *crusta*, a shell) are mainly aquatic creatures and are abundant in both the sea and fresh waters. The groups of smaller crustaceans are often referred to collectively as the Entomostraca (Gr. *entomon*, an insect; *ostrakon*, a shell), to distinguish them from the Malacostraca (Gr. *malakos*, soft), a division which contains the higher forms, but the term, although perhaps convenient, is not a scientific basis of classification, for the lower groups differ from each other as much as each of them does from the Malacostraca.

Sub-class *BRANCHIOPODA* (Gr. *branchion*, a gill; *podos*, a foot).

Order ANOSTRACA

The branchiopods, or gill-footed crustaceans, owe their names to the flattened swimming appendages on the thorax which serve as a form of gill to assist in breathing. This feature is well seen in the fairy shrimp, *Chirocephalus diaphanus*.

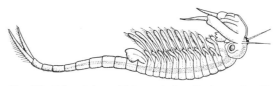

Fig. 56 Fairy shrimp, *Chirocephalus diaphanus* (× 3)

This lovely creature, which is about 25 mm in length, has eleven pairs of leaf-like limbs on its thorax, and their vibrations not only propel it along but also keep up a constant circulation of water over the body, bringing fresh supplies of oxygen.

The large number of distinct segments of the body, and the similarity of their appendages, indicate that the fairy shrimp is one of the more primitive crustaceans. The head is distinct and bears two large compound eyes on movable stalks and two pairs of antennae, the front pair thin and the rear ones broad. It is by the antennae that the sexes can be distinguished, for whereas in the female the rear pair are triangular in shape, in the male they are complicated lobed organs which serve to grasp the female during pairing. The body tapers off posteriorly to end in a forked tail.

Fairy shrimps are very active animals, incessantly in motion and usually swimming on their backs, although they do sometimes turn over and sweep the bottom. The transparency of the body and its iridescence, which have earned them their common name, make them particularly beautiful creatures to observe as they swim gracefully through the water. Food consists of microscopic algae or animals which are strained out of the water by the limbs and then conveyed to the mouth. The straight alimentary canal is plainly visible inside the transparent body, and the nature of the food can usually be determined by the colour of the contents.

The places in which to find fairy shrimps are small temporary pools that dry up in the summer, and the creatures have even been found in deep cart-ruts. The eggs, which are deposited in a brood-pouch situated just behind the leaf-like appendages of the female, sink into the mud and survive long periods of drought, until water again fills the pool and they are able to develop. Some species of branchiopods seem, in fact, unable to thrive unless the eggs undergo a period of drought.

The life-history of these creatures lends itself to their dispersal by birds and possibly wind, and it is characteristic of some of the members of this group to appear suddenly in new localities, multiply rapidly for a season or two and then disappear, possibly because their numbers have exhausted the available food-supply in the restricted environment.

From the egg hatches what is called a *nauplius* type of larva, a creature unlike the parents, with two pairs of long antennae and another pair of long appendages which are used in swimming.

Order NOTOSTRACA

A large shell or *carapace* is a common possession of many crustaceans, and typical ones can be seen on crabs and lobsters. Although the fairy shrimp is not so endowed, the related creature known as apus, or *Triops cancriformis*, has a large part of its body covered with a shield-shaped carapace, from beneath which projects the hinder part of the trunk ending in two tails.

This strange creature, which is larger than the fairy shrimp, is very rare in Britain, but has been recorded from a number of localities as widely separated as Kent, Kirkcudbrightshire, Gloucestershire and Hampshire. Like the fairy

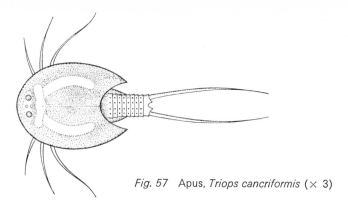

Fig. 57   Apus, *Triops cancriformis* (× 3)

shrimp, it frequents temporary pools, and the last place in which it was found was a shallow grass-bottomed pool which was waterless in dry weather. Only the females are generally found in this country, but testis lobes are usually seen on the animal's ovary and *Triops* would appear to be hermaphrodite.

Order CLADOCERA

Probably the best known and commonest of the branchiopods are those belonging to the order Cladocera (Gr. *klados*, a branch; *keras*, a horn), in reference to their branched antennae. Two distinct groups of cladocerans occur, Calyptomera (Gr. *kalyptos*, covered; *meros*, a part), which includes the more typical ones such as *Daphnia*, largely enclosed in a shell or *carapace*, and Gymnomera (Gr. *gymnos*, naked, lightly clad), a few strange forms in which the shell only covers a pouch in which the eggs are carried.

Members of the first group, commonly called 'water fleas', abound in most stretches of water, some forming a large part of the plankton of lakes; others are found in ponds and ditches, often in incredible numbers. Apart from the head, which bears the two pairs of antennae, the rest of the body and limbs are enclosed in what appears to be a shell made of two parts or valves, but is, in fact, all in one piece and folded over the back. The transparency of the shell often enables the whole internal structure of the animal to be observed, and the limbs are also clearly visible. The beating of the heart, the passage of food down the gut, the movement of the muscles and the circulation of the blood can all be studied under a low power of the microscope in a specimen whose activity is restrained in a 'live-box'. The large, single but compound black eye is a very noticeable feature of the water flea, and the muscles which control it can be seen.

The movements through the water of *Daphnia*, the best known of the cladocerans, consist of a series of jerks or hops, and no doubt it is this characteristic which has earned these crustaceans their common name of 'water fleas', although they are, of course, not related to the true fleas. The second pair of antennae are the main swimming organs, and the vigorous strokes of these produce the familiar upward jumps. At each pause in the motion, the creatures sink back almost to their original position, so that it seems as if their existence is one long struggle to keep up in the water. Most cladocerans are sensitive to light intensities, and the particular level at which they will be found in the water depends on the strength of light at the time.

Minute free-floating algae and bacteria form the main food of cladocerans, and these are strained out of the water current passing through the shell by the bristles of the legs, and then passed to the mouth. During summer, when such food is plentiful, the creatures multiply rapidly, and since many of the larger creatures, including fish, eat them with avidity, their function as converters of plant matter into animal food is of the greatest importance in the economy of fresh waters.

During the greater part of the year only female water-fleas are found, and they lay eggs which do not require fertilizing. These so-called 'summer eggs' are laid, perhaps twenty or more at a time, and kept in a space just under the top margin of the shell called the brood-pouch. Here, in a day or so, the eggs develop into young water-fleas, miniature replicas of the adult. At certain times, which seem to coincide with unfavourable conditions in the water, such as the approach of winter, or drought, or the absence of food, some of the eggs hatch into males, and subsequently small numbers of a special type of egg which needs fertilization are then laid by the females. To accommodate these the brood-pouch develops thickened walls enclosing, as in a little box, the two or three 'winter eggs'. (Plate 26, 2) From their resemblance to a saddle, these special egg compartments are called *ephippia* (Gr. *ephippos*, mounted as on a horse). When the female moults, her ephippium breaks loose and either floats or sinks to the bottom (as happens in the case of *D. magna*), until conditions once again favour the development of the eggs. From these, females hatch out to start the cycle anew. Many ephippia are no doubt carried by birds, attached to their feathers or in the mud of their feet, to other localities.

Sida crystallina is a common form in clear ponds and at the edges of lakes, usually attached to water-plants by a gland at the back of the head. When captured, it frequently adheres to the net or tube in the same way. *Holopedium gibberum* is a species found only in the plankton of lakes, swimming always on its back. Its body is flattened from side to side, and the rear part projects outside the shell. The animal is surrounded by a clear, gelatinous mantle; bright red or blue specimens are sometimes seen.

The many species of *Daphnia* sometimes occur in incredible numbers in suitable ponds. Frequently reddish or brownish patches in the water will prove on close examination to be huge swarms of the creatures. In badly oxygenated waters *Daphnia*, in common with other cladocerans, usually assumes quite a reddish colour, due to an increase of haemoglobin in its body which assists the creatures in respiration by making the most of what available oxygen there is.

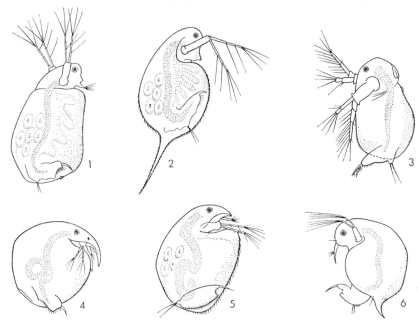

Fig. 58 Crustaceans: 1 *Simocephalus vetulus,* 2·5 mm. 2 *Daphnia hyalina,* 2 mm. 3 *Sida crystallina,* 3 mm. 4 *Chydorus ovalis,* 0·6 mm. 5 *Eurycercus lamellatus,* 4 mm. 6 *Bosmina coregoni,* 1 mm

Daphnia magna is the largest species, females being up to 5 mm long, and *Daphnia obtusa* is probably the commonest in ponds, in most parts of the country. The closely related genus, *Simocephalus,* can be distinguished by its rather squarer body and by its absence of a tail spine. *Simocephalus vetulus* is the commonest species, and is found among water-plants in ponds all over the country. *Eurycercus lamellatus*, another large species, up to 4 mm long, is common and widely distributed, living among water-plants. When caught, it seems to become trapped in the surface film very easily.

The species of *Bosmina* are all small cladocerans, the lagest being about 1 mm in length. The first pair of antennae are not movable and project from the head rather like a trunk. *B. coregoni* occurs in the open waters of ponds and lakes except in south and eastern England. *B. longirostris* is commoner in these areas and less frequent in the north.

Anchistropus emarginatus, a small cladoceran less than half a millimetre long, is sometimes free-swimming, but more usually found as a parasite on hydra (page 86), attached by the first trunk limbs and ventral part of the shell while it moves over the surface of its host.

The cladocerans in the Gymnomera group are quite different both in appear-

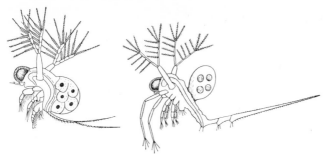

Fig. 59 Crustaceans: *Left: Polyphemus pediculus,* 2 mm.
Right: Bythotrephes longimanus, 2–3 mm

ance and habits from those in Calyptomera. Their bodies and limbs are not enclosed in a shell, but a minute shell is modified to form the egg-pouch.

They are all predacious creatures, feeding on Protozoa, rotifers and small Crustacea. *Polyphemus pediculus* has a short body about 2 mm in length, the most noticeable feature being a huge black eye which, even when the creature itself cannot be seen, appears like a black speck in the water. A long spine projects from the rear end of the body. *Polyphemus* is abundant near the margins of lakes and clear pools, particularly in late summer.

Bythotrephes longimanus, belonging to the same family as *Polyphemus,* but distinguished by longer appendages, occurs in the plankton of northern lakes. *Leptodora kindti* (= *Leptodora hyalina*) is one of the most transparent creatures known, and if a tube containing them is held to the light, only the dark eyes and contents of the digestive tracts will be visible. Seen under dark-ground illumination through a microscope, however, it is a beautiful object. *Leptodora* is the largest cladoceran, the females reaching a length of about 10 mm. The body is elongated and cylindrical, with a greatly extended head bearing the single, large eye. The fore part of the body carries the six pairs of limbs. The second pair of antennae are very large and strong, and serve as most efficient swimming organs. On the back the egg-pouch, formed from what little shell the creature has, projects as a loose bag, attached to the body only at its narrow mouth end. Unique among cladocerans, the winter eggs of *Leptodora* give rise to a nauplius larva, although the normal summer eggs develop directly into a miniature of the adult. *Leptodora* occurs only in the plankton of large stretches of water, rising near the surface at night. (Plate **26, 1**)

Sub-class *OSTRACODA* (Gr. *ostrakon,* a shell)

The name of this group is very apt, for on the first sight of an ostracod there seems to be little but a bean-shaped shell. The entire body of the animal is enclosed in it with the exception of the antennae and a pair of slender legs at the rear end. When the creature is disturbed or alarmed, even these parts are quickly withdrawn and the shell closed tightly, and, thus deprived of any means of swimming, the animal sinks to the bottom of the water. Although the shell is all in one piece, it is functionally bivalve, rather like that of a mussel. It serves, not merely as an armour-plating as in other crustacea but, as one writer has put it, also as 'a portable funk-hole of maximum efficiency'.

In most species the shell is too opaque to make out much detail of the structure of the body within, but there are, in addition to the two pairs of antennae, three pairs of limbs which are modified into jaws and two pairs of legs, being used not only for walking, but as cleansing organs. A pointed structure called a *furca*

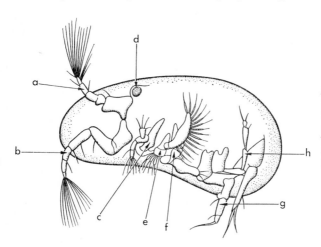

Fig. 60 Structure of an ostracod (× 60). a. antennule, b. antenna, c. mandible, d. eye, e. 1st maxilla, f. 2nd maxilla, g. 1st thoracic limb, h. 2nd thoracic limb

is sometimes protruded out of the shell. Near the top of the back of the animal is the single, simple eye.

Ostracods scuttle actively about in the water, sometimes swimming by means of backwards and forwards movements of their antennae, and sometimes scrambling among water-plants, using their first pair of legs. They feed mainly on decaying organic matter, pulling the larger pieces into their shells; smaller particles are drawn inside by currents caused by the movements of the limbs.

In some species both sexes occur regularly, but in others no males have been found and presumably only parthenogenetic eggs are laid. Ostracods do not follow the usual crustacean habit of carrying their eggs about, but deposit them in spring in little orange-coloured clusters on water-plants. The eggs will withstand drying-up, and can survive for years in this state. An instance is on record of eggs of *Cypria* hatching after being kept in dried mud for about thirty years. From the egg, nauplius larvae with rudimentary shells hatch out.

Cypria is probably the best-known genus of ostracods, sometimes occurring in immense numbers in weedy ponds. The largest European ostracod is about 3 mm long, but most are only about 1·5 mm. The colouring is very variable even among members of the same species, and seems to depend on the type of habitat in which they happen to be living. Brown and yellow are perhaps the most usual colours, but green specimens are frequently found. (Plate 27).

Sub-class COPEPODA (Gr. *kope*, handle, oar; *podos*, a foot)

The copepods form a large group, most of the members of which are free-swimming, although some, mainly marine, are parasitic on other creatures. The free-swimming forms are very common in almost all stretches of fresh water, and may be found even in the winter months. They are readily distinguished from the other freshwater Crustacea we have been considering by the absence of a shell and the shape of their body (page 152), which is clearly divided into a cephalothorax and an abdomen, each composed of distinct segments. On the head are a pair of long antennules, in some species bent in the male to hold the female, and a pair of shorter antennae, often invisible when one is looking at the back of the animal, which is the usual position. On the underside of the body, similarly hidden from view, are numerous other pairs of limbs, some of which are used for swimming and others for grasping food. The limbs and tail appendages bear stiff bristles. In the centre of the head is a single eye, usually of a red or black colour, and *Cyclops*, the best-known genus, is so named because of this feature, in allusion to the giants of Greek mythology whose single eye was situated in the centre of their foreheads.

The body colouring of copepods is often very striking, sometimes being bright red, green or even blue, but many species are almost colourless and transparent. The different species vary somewhat in their food requirements, large cyclopoids being carnivores and others eating algae and detritus. Reserve food is sometimes stored in their bodies in the form of oil globules.

Probably to the average student of freshwater life, the most noticeable feature of the commoner species of copepods is the manner in which the females carry their eggs in one or more large masses, often called the egg-sacs, attached by a slender neck to the rear part of their bodies. As in other freshwater crustaceans, two kinds of eggs are laid by some species, one in which development is completed in about ten days, and the other a resting type capable of tiding the creatures over an unfavourable period.

Male copepods do occur, usually being smaller than the females. In some species one or more antennae are modified for clasping the females when pairing. The sperm is transferred to the female in a receptacle called a spermatophore which, once having been attached, remains in position until the female dies.

The little creature hatching from the egg is always a nauplius larva, which takes about three weeks or a month to develop into a fully adult copepod. It is a simple matter to obtain the nauplius larvae of, say, *Cyclops* for study if a female carrying eggs is pipetted into a very small container, such as a watch-glass, filled with pond water to ensure that there is some food available. A close observation

L

of the *Cyclops* under a lens will show when the eggs have hatched, and it should be a fairly simple matter to locate the nauplii in the small quantity of water under a medium power of the microscope. They are, however, sometimes a little difficult to see at first as they are almost transparent, but their dark eye will usually give away their presence. (Plate **27**, 2)

Like cladocerans, free-swimming copepods are an important item in the diet of fish and other creatures. They are also, as we have seen, intermediate hosts for parasitic worms such as the tapeworm, *Schistocephalus*, which depends for its development on the copepod being eaten by a fish.

The parasitic copepods in their adult stages are often unlike the free-swimming forms in appearance, but their nauplius larvae are similar, so that we may assume that the strange shape possessed by many of them is merely a form of degeneration which has come about as a result of their mode of life. They are mainly parasitic on fish, living in the gill region and sucking the blood. Like all parasites

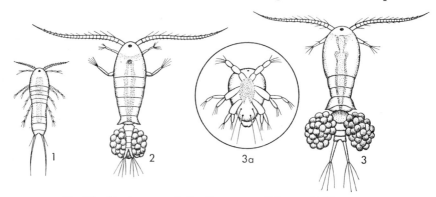

Fig. 61   Crustaceans: 1 *Canthocampus*, 0·5 mm. 2 *Diaptomus*, 1–2·5 mm. 3 *Cyclops*, 2 mm. 3a nauplius larva of *Cyclops* (× 75)

they are prolific creatures, and the egg-sacs of the females are larger than those of the free-swimming kinds. In most species the females alone are parasitic and the males are free-swimming, but in others both sexes have adopted a parasitic existence.

There are three main orders of the freshwater copepods, and their characteristics, together with a brief mention of a few representative members, are given below.

CALANOIDA   In this group the antennules are exceptionally long, usually nearly as long as the whole body of the creature, with 17 to 25 segments. The most familiar members of this group are the various species of *Diaptomus*, which are about 2·5 mm in length, and are identifiable by their gliding movements in contrast to the jerky swimming movements of cyclopoids. *D. castor* is plentiful in ponds and ditches during the winter months only, but *D. gracilis* and *D. vulgaris* occur also during summer. *Diaptomus* obtains its food by sieving floating particles out of the water. Only one egg-sac is carried by the female and this is usually reddish in colour.

CYCLOPOIDA   The majority of the species in this group belong to the genus *Cyclops* and are common in almost every type of still water. The fore-part of the body is broader than in *Diaptomus* and clearly separated from the narrower abdomen. The antennules are not so long as in the latter creature, and consist only of between 6 and 17 segments. The females carry two egg-sacs and, although the animals are never more than about 3 mm in length, it is quite easy to see these sacs with the naked eye. (Plate **26, 3**)

Cyclops actively seizes food particles (in contrast to *Diaptomus* which merely strains them from the water), and the larger species devour live animals.

The main swimming appendages are on the underside of the thorax and, when disturbed, the creatures can move quickly with a characteristic jerky action.

Some of the Cyclopoida are parasitic or semi-parasitic on fish. *Ergasilus sieboldi*, which is common on many species of freshwater fish in continental Europe, has been recorded in Britain on trout, attached to the gills. Only the females are parasitic, the males living a free-swimming existence.

Thersitina is parasitic on sticklebacks living in brackish water. Again it is only the females that are found on the gill-covers of the fish. Both species resemble a typical copepod, but the egg-clusters are exceptionally large.

Although a marine species, the so-called sea-louse, *Lepeophtheirus salmonis*, is found on salmon in rivers soon after the fish have returned from the sea to breed. They are usually near the anal region of the fish and both males and females occur, the former being about 5 mm long and the latter 15 mm. They do not survive long in fresh water.

The gill maggot, *Salminicola salmonea*, found on salmon is another parasitic copepod which breeds in fresh water but survives after the fish enters the sea.

HARPACTICOIDA   Most members of the third group of copepods are much smaller animals than those in the preceding two groups, many of them being under 1 mm long. They have small antennules, which are never composed of more than ten segments. There is no clear distinction between the cephalothorax and the abdomen, so that the animals are rather worm-like in appearance, a resemblance which is further accentuated by the way they creep over the surface of water-plants or the bottom of the pond. They are not free-swimmers like the other two groups.

There are about thirty-two true freshwater species of harpacticoids in Britain, *Canthocamptus staphylinus* being perhaps the commonest. It is most abundant in the winter months, at which time the females will be carrying their single egg-sac. The male sperms are also carried externally by the female in a spermatophore. Nauplius larvae hatch from the eggs.

Sub-class *BRANCHIURA* (Gr. *branchion*, a gill; *oura*, a tail)

The fish-lice (Argulidae) resemble the copepods in structure, but they are now generally considered to be a distinct group. They are, of course, not lice at all, for a louse is an insect, but their habit of remaining attached to the outside of fish and sucking their blood has given rise to the name.

They are odd-looking creatures, but beautifully adapted for remaining firmly attached to the body of their host, no matter how swiftly it swims through the water. The huge suckers on their under-surface, resembling in appearance big round eyes, enable them to take a firm grip of the fish, and their extreme flatness ensures that they offer practically no resistance to the water, so that there is little danger to them of being swept off. In addition, they have backward-pointing spines which are further aids to maintaining their position.

Between the suckers is a poison spine and behind them a proboscis through which the food, rasped from the body of the fish, is taken. Both the antennules and antennae are considerably reduced in size, but the compound eyes just behind them are conspicuous. There are four pairs of feathery swimming limbs which are constantly jerked backwards and forwards even when the parasite is attached to a fish. A very small pair of limbs is situated in front of them, just behind the suckers. The extreme thinness of the body and its transparency enable most of the creature's structure to be made out whether it is viewed from the front or back. (Plate 27, 3)

Fish-lice are commoner than is perhaps realized, and have been found on carp, bream, tench, dace, pike, perch, trout, minnows and sticklebacks. They may be attached to almost any part of their host, although they are rarely seen on the dorsal region. Many may be found on a single fish, and the effect on the unfortunate victim can be well imagined. Sometimes a fish will be seen repeatedly dashing itself against a stone in an effort to dislodge its unwelcome passengers.

On the death of their host, or even at other times, fish-lice swim freely in the water in search of another victim, and at breeding time the females swim round until they find a suitable stone or other solid object on which to attach their eggs. The eggs are oval in shape and white when first laid, usually in July or August. Over a hundred may be deposited by a single female, the mass being covered with a gelatinous envelope. The larvae hatch in about a month.

Argulus foliaceus is the commonest species, and the females, which are larger and more frequently seen, are about 8 mm in length. The colour varies from light green to brown and probably depends on whether the particular specimen being examined is well-fed or not. Another species, *A. coregoni*, is larger, but is not found so frequently in this country. The specimen illustrated on Plate 27 was taken in Somerset on trout.

Sub-class *MALACOSTRACA* (Gr. *malakos*, soft; *ostrakon*, a shell)

The higher Crustacea which are included in this class all share one feature: their body is made up of twenty segments, each one of which, with the exception of the tail segment, has a pair of appendages. The head accounts for six of these segments, the thorax eight and the abdomen the rest; a carapace encloses the thorax at the sides. Although there are many different kinds, only five orders have freshwater representatives. Most are much larger than the crustaceans considered so far.

BATHYNELLACEA. A primitive crustacean *Bathynella natans*, only about 1·5 mm long, is found in subterranean waters such as those in caves and wells. It may, however, have a more widespread distribution than has been realized, living in ground water between the interstices of gravels. It is colourless and without eyes; the segments of the thorax, except the last two, and the first and last segments of the abdomen, have branched appendages.

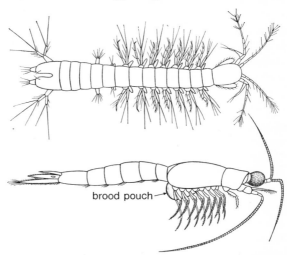

Fig. 62 Top: Bathynella natans, 1 mm. *Below: Mysis relicta*, 15 mm

brood pouch

MYSIDACEA. *Mysis relicta*, a prawn-like crustacean, about 15 mm long, is the only British freshwater representative of this order. It is believed to be an estuarine species which became widely distributed by the last ice age and then isolated in basins of fresh water when the ice retreated. In England it occurred in Ennerdale, in the Lake District, but has not been seen there for some years. In Ireland it has been recorded from Loughs Corrib, Ree, Erne, Derg and Neagh, and also from the rivers Shannon and Bann.

ISOPODA (Gr. *isos*, equal, similar; *podos*, a foot). The best-known members of this order are the familiar water-lice, hog slaters or water slaters, *Asellus*, which are common, often extremely so, in weedy ponds and streams, clambering among the plants or scuttering on tip-toe on the bottom. They are between 18 mm and

25 mm in length, with a flattened body and, as the name of the order implies, with legs all of the same type. To see the appendages it is necessary to turn a specimen on its back, when antennules and antennae, four pairs of mouth-parts, seven pairs of legs on the thorax, five broad plate-like gills, and two uropods on the abdomen will be observed. The compound eyes are sessile—that is, not borne on stalks. (Plate 28)

Water-lice are largely scavengers and feed on all kinds of decaying organic matter, but also eat algae. If specimens are examined in spring, the females will be found to be carrying their whitish mass of eggs on the underside of the body near the head. The eggs are held by flat plates arising from the bases of the first four pairs of legs. The young resemble the adults when they hatch, and remain attached to their mother for a time.

If a water-louse is held in a live-box and a portion of the antenna examined under a medium power of the microscope, the individual blood corpuscles will be seen rolling along like small marbles.

Asellus aquaticus and *A. meridianus* are the commonest species and are widely distributed. They have undoubtedly been confused in the past. *A. aquaticus* is the larger of the two and darker in colour. The antennae, too, in *A. aquaticus* are almost as long as the body, whereas in *A. meridianus* they are only two-thirds the body length. It is an interesting fact that in winter *A. aquaticus* is usually found in pairs, the male carrying the female under him, but this habit has not been observed in the other species. It has been suggested that this may be due to the large excess of males of *A. meridianus* which seems to exist, the sexes of *aquaticus* occurring in almost equal numbers. *Asellus cavaticus* lives in subterranean waters in caves and wells and has been recorded from southern England and South Wales.

AMPHIPODA (Gr. *amphi*, both, in reference to the two types of thoracic limbs). A typical amphipod crustacean is flattened from side to side (thus differing from an isopod, which is flattened from above downwards), and the body when at rest is curved round to form an arc of a circle. It usually swims or 'scuds' along on its side, often with surprising speed, and when it is moving the hind part of the body is straightened out, only to contract again suddenly into its normal curved position as soon as the creature stops.

The freshwater shrimps, *Gammarus*, are good examples of this order, and are very common creatures in streams and rivers all over the country. They are usually found under stones or on the soft surface of the mud, and when disturbed scud rapidly away to shelter. The name is unfortunate, for they are not true shrimps.

The males are about 25 mm in length and the females slightly smaller. Frequently they will be found swimming together, the male carrying the female under his body. The colour is usually lightish brown, but may vary. The head bears the compound, sessile eyes and the two pairs of long antennae. The gills are situated at the base of the first four pairs of thoracic legs, and to keep these supplied with fresh currents of water, the first three pairs of limbs on the abdomen are frequently vibrated when the creature is at rest. The thoracic limbs, three pairs of which are reflexed backwards over the carapace when the animal is at rest, and the first three pairs on the abdomen are used in swimming; the last three pairs on the abdomen, which are directed backwards, are used to kick the ground when scrambling around. (Plate 28)

Freshwater shrimps are largely scavengers, feeding on decaying organic matter, but they are not averse to attacking and devouring a small living creature, and have even been observed indulging in cannibalism.

The eggs are carried by the female in a brood-pouch under the thorax, and here, too, the young may be found for a few days after they have hatched. There is no free larval stage, the young resembling their parents. Occasionally, the whole family may be found swimming about together, the male carrying the female with her attached eggs or young.

Three species of *Gammarus* are found in our fresh waters and a number occur in brackish waters. *G. pulex* is the commonest, except in Ireland where *G. duebeni*, elsewhere mainly a brackish species, is found abundantly in fresh water. *G. lacustris* has been recorded from Scottish lochs, and lakes in northern England, Wales and Ireland. *G. tigrinus* is primarily a brackish water species, but is found in fresh waters where industrial processes have brought about conditions of high salinity, as in the north and west Midlands.

Crangonyx pseudogracilis, a species which was introduced from North America about 1930, is now widely distributed in Britain. It resembles *Gammarus* but is only about 15 mm long when adult, and is often bluish-green in colour. When moving in the water it does so in an upright position, whereas *Gammarus* swims on its side.

Niphargus, a slender, colourless amphipod about 12 mm in length, is found in underground waters and wells in southern England, Wales and Ireland. Four species and two subspecies have been recorded in Britain.

DECAPODA (Gr. *deka*, ten). The last order of the crustaceans to be considered contains the largest, most highly organized and possibly the most interesting of them all—the freshwater crayfish, a creature resembling a small lobster in appearance. Although small compared with its marine relatives, it is a monster compared with the tiny creatures we have been considering in this chapter, for even average adult specimens are 10 cm or more in length and much larger ones are occasionally found. The females are smaller than the males. The colouring is usually dark brownish, but greenish or yellowish forms occur.

Crayfish need well-oxygenated water and live only in streams or rivers where this condition obtains. They thrive best in 'hard' waters, and are therefore found more often in streams in limestone or chalky districts. Only rarely will they be seen moving about in the day-time, for they are largely nocturnal creatures, hiding under large stones or in burrows in the bank during the day.

The head and thorax of a crayfish are fused together and covered by a shield-like carapace. The compound eyes are borne on stalks, and the five pairs of appendages on the head region comprise: a pair of small antennules well supplied with sense organs; a pair of long antennae used as organs of touch and having excretory openings at their base; the strong crunching jaws; and finally two pairs of maxillae or accessory jaws, the second of which is used for maintaining a current of water over the twenty pairs of feathery blood-gills situated at the base of the thoracic limbs.

The thorax carries three pairs of appendages, which help in passing food to the jaws, and also the five pairs of legs from which the name of the order is derived. The first pair of these are the strong pincers which are used in combat or for grasping prey, and the remaining four pairs are the walking legs. Ordinarily,

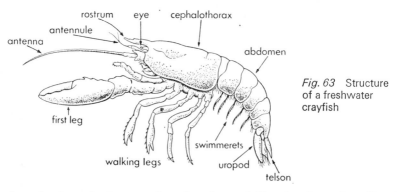

Fig. 63 Structure of a freshwater crayfish

the animal crawls along on these in a forward direction, but when disturbed it moves backwards at great speed, using its fan-shaped tail as a paddle.

The abdomen is very clearly divided into segments, the first of which carries a pair of appendages which in the male are deeply grooved and used for passing the sperms to the female. The next four pairs are two-bladed 'swimmerets', which assist in swimming and are also used by the female for carrying the eggs. (Plate 29)

Crayfish are mainly carnivorous, capturing and devouring almost any kind of smaller creatures such as insect larvae, water-snails and tadpoles, but occasionally taking vegetable food. When kept in an aquarium they feed readily on worms, but if food is inadequate, they have been known to turn cannibal. If a number are kept together they will frequently eat the moulted cuticles of their fellow inmates, possibly to obtain the lime salts contained in them.

Pairing takes place in late autumn. The male overturns the female and liberates a fertilizing liquid on her abdomen, where it adheres. Shortly after this, in the seclusion of her burrow, the female lays her hundred or more pinkish eggs, attaching them to the bristles of her swimmerets, where they come into contact with the spermatazoa of the male and are fertilized. The eggs do not hatch until spring, and the tiny transparent young crayfish, almost miniature replicas of the adults, remain attached to the female for some time afterwards, grasping the swimmerets with their hooked pincers.

The scientific name of our freshwater crayfish is usually given as *Astacus fluviatilis*, but this is the name first given to a larger continental species which is not native to Britain. In this species the underside of the pincers is usually reddish in colour, and on this account it is called the 'red-claw'. It is an important article of commerce, and is reared on a large scale on the Continent, particularly in France, in crayfish farms. Considerable quantities were formerly imported into Britain as an article of diet and as a zoological type for study in schools and colleges, and several attempts to cultivate the creature in England have been made in the past. From one such venture at Nuneham Park, near Oxford, many years ago, abortive in itself, some specimens escaped and found their way into the streams and backwaters of the Thames where, except for degeneration in size, they prospered. No doubt similar attempts to cultivate the creatures have resulted in 'escapes' in other parts of the country, so that the 'red-claw' may be found in a wild state in Britain.

The true native species is, however, *Astacus pallipes*, the 'white-claw', which is smaller than normal specimens of the continental species, and thus not so acceptable as an article of food, although large numbers certainly found their way to the table in the past. A favourite way of capturing them was to tie a piece of meat to the centre of a hoop-net which was then lowered at night on to the bottom of the stream and raised at intervals with the crayfish firmly attached to the bait by their pincers.

This interesting creature is apparently becoming rarer, and its virtual disappearance from localities where it was formerly abundant has been attributed to an epidemic disease possibly caused by a fungus, *Aphanomyces astaci*. River pollution and possibly surface drainage from tarred roads running into their streams may also be partly responsible for their decrease, for they are known to be susceptible to the presence of such impurities in the water.

14: INSECTS, PART ONE

The insects of ponds and streams are perhaps the most noticeable members of the freshwater fauna and it may seem strange, therefore, to say that they are not true aquatic creatures at all, but have taken to life in the water at some late stage in their evolution.

With the exception of the bugs and beetles and a relatively unimportant group of ichneumon flies, the adult aquatic insects are not found in the water, and it is with the immature stages that we are more concerned.

The characteristic features of a typical adult insect are the segmented body, divided into the head, the thorax of three segments of which the last two usually bear wings, and the abdomen of about ten segments; the possession of three pairs of legs; a pair of feelers or antennae which are touch and smelling organs; two kinds of eyes—relatively simple ones called ocelli placed on the top of the head, and large compound eyes which may contain many thousands of individual facets.

Less obvious, but important to the entomologist as aids to identification, are the mouth-parts, which vary somewhat between different kinds of insects, but usually consist of the *mandibles*, a pair of strong biting or chewing jaws; the *maxillae*, a pair of smaller jaws behind them provided with *maxillary palps* which are for the purpose of testing food; and the *labium*, which consists of a pair of jaws joined together to form a lower lip. The labium is also provided with sensory palps.

The outer covering or 'skin' of insects is, like that of other arthropods, composed of chitin secreted by the underlying skin-cells, and is similarly inelastic. Insects, however, do not change size when once they have reached the adult stage, and immature insects during their development get over the difficulty by moulting their covering at intervals when it becomes too small and secreting another.

The stages through which an insect such as a butterfly passes during its development are known to everyone—the egg; caterpillar (larva), devouring all before it; the pupa, usually an immobile stage; and finally the adult winged insect (imago). All other insects except some very primitive ones pass through a series of similar stages, but in some a resting pupal stage is omitted, and the immature insect passes gradually into a creature closely resembling the adult.

During the latter phases of this development wing buds appear on the back of the creature, and with a final moult the fully adult winged insect comes into existence. Such a metamorphosis is termed *incomplete* and the immature insect is usually called a nymph, the usual development through larva and pupa being described as a *complete* metamorphosis.

The breathing of most insects is effected by taking in air at air-holes or *spiracles* along their bodies (Plate **30,** 1), and passing it through a system of branching tracheal tubes direct to every part of the body requiring it. In immature stages, and in less active or small adult insects, the air is taken in and expelled by simple diffusion, but in most others, active respiratory movements are made, expiration being effected by the contraction of the body through the action of the abdominal muscles, and inspiration as the body segments regain their original shape.

This method of breathing is simple and efficient for insects living on land, but when we find that most insects living in water depend on it also, we are left in no doubt that they are not truly aquatic but are creatures that have at some time of their development left the land and taken to this alien environment. With the typical versatility of their kind, however, water-insects have all perfected their own ways of overcoming the disadvantages of their terrestrial origin, as will be seen when the individual species are being dealt with, but most of these methods can only be considered as makeshift devices, although efficient ones, to enable them to carry air from the surface down into the water for use there.

In some larvae the modification to an aquatic existence has progressed further and they possess gills of a kind. These take several forms, but are basically thin membraneous extensions of the body chitin, traversed by minute branches of the tracheae. The interchange of gases involved in the breathing process takes place by diffusion through the thin walls of the gill. In very young larvae, and even in older larvae possessing very thin chitinous coverings, this interchange of gases can take place even through the chitinous covering of the body itself with or without tracheal gills.

Many adult aquatic insects carry a bubble or film of air trapped in a velvety pile of unwettable fine hairs on some part of their body. Their spiracles open into it and gases are exchanged between both the trachea and the bubble, and the bubble and the surrounding water. The gas store therefore acts in two ways: as a supply on which the insect can draw by means of its spiracles while it is submerged, and as a physical gill which exchanges gases with the surrounding water. In the latter role the inert gas, nitrogen, plays an important part. As oxygen is used by the insect, and its expired carbon dioxide dissolves into the water, the bubble shrinks and the pressure of nitrogen in it is increased; this causes nitrogen to diffuse out and more oxygen to diffuse in from the surrounding water. In time the bubble decreases, through the slow diffusion of nitrogen outwards, to a point when the insect has to rise to the surface to renew it.

In this compressible kind of bubble that occurs in insects such as beetles and bugs, which periodically come to the surface to renew their air supply, the oxygen diffusing into it is often more important than the quantity present in it initially. Experimental work has shown that in a water-boatman, *Notonecta*, it may provide more than thirteen times as much as the insect could obtain from its original air store.

A second kind of incompressible physical gill is found in some insects which live permanently below water. Gas is trapped as an extremely thin film of constant volume among very short, tightly packed hairs which are, moreover, sometimes bent over at their tips. In the bug *Aphelocheirus* (p. 178) there are more than two million hairs per square millimetre. They are so closely set that the gas store is not lost when pressure inside decreases. Such a gas store is called a *plastron*, and it acts as a permanent physical gill, enabling its possessor to remain below the surface indefinitely, obtaining all the oxygen it needs from the surrounding water. In addition to *Aphelocheirus*, beetles of the genera *Elmis* and *Dryops* (p. 192), and the parasitic ichneumon larvae, *Agriotypus* (p. 210), have such stores.

Of the twenty-nine orders into which insects are divided, the following eleven have aquatic representatives:

Sub-class: *APTERYGOTA*—completely wingless and without a metamorphosis.
  1 COLLEMBOLA   Two of the springtails live on the surface of the water. They are of small size, without wings, and provided with a forked tail appendage with which they are able to spring into the air.
Sub-class: *PTERYGOTA*—nearly always with two pairs of wings and undergoing a metamorphosis.

Section: **EXOPTERYGOTA**—metamorphosis gradual or incomplete.
  2 EPHEMEROPTERA   Mayflies. All exclusively aquatic in immature stages; membraneous wings of which the fore pair are larger; long tail filaments; mouth-parts of adult vestigial.
  3 ODONATA   Dragonflies. All exclusively aquatic in immature stages; wings large and membraneous; biting mouth-parts.
  4 PLECOPTERA   Stoneflies. All exclusively aquatic in immature stages; membraneous wings; biting mouth-parts; antennae and tail filaments long.
  5 HEMIPTERA   Bugs. Some of the sub-order Heteroptera aquatic in all stages; typically two pairs of wings, the first pair hardened and modified as wing-cases; piercing and sucking mouth-parts. Aquatic species belong to the sub-order Heteroptera.

Section: **ENDOPTERYGOTA**—metamorphosis complete.
  6 NEUROPTERA   Sub-order Megaloptera: alder-flies; larvae aquatic; adults with large, similar wings forming a broad roof over the back when at rest. Sub-order Planipennia: spongilla- or sponge-flies, *Sisyra* and *Osmylus*. Wings similar and membraneous. Biting mouth-parts. Larvae aquatic.

7 COLEOPTERA   Beetles. Many species aquatic in both immature and adult stages, although the pupation period is usually spent out of the water. Two pairs of wings, the fore pair (elytra) being hard, stiff cases for the rear pair; biting mouth-parts.

8 TRICHOPTERA   Caddis flies. Larvae and pupae aquatic. Wings and body covered with hairs; larvae with pair of hooked structures at tail end, and usually live in cases.

9 LEPIDOPTERA   Butterflies and moths. Some moths of the family Pyralidae pass their immature stages under water. Two pairs of large wings covered with scales, long, coiled, sucking mouth-parts (proboscis).

10 DIPTERA   Two-winged flies—many species passing immature stages in water. Fore wings only, hind pair being represented by a pair of small plate or club-like organs (halteres); mouth-parts either piercing and sucking, or modified for sucking up surface liquids.

11 HYMENOPTERA   Ants, bees, wasps, etc. Some small ichneumon flies, *Agriotypus*, *Caraphractus* (= *Polynema*), *Prestwichia*, are parasitic on other aquatic insects; and the immature stages of the first-named, and the adults of the two last-named genera live under water.

Springtails: Order COLLEMBOLA

A few insects are regarded as aquatic even though they rarely, if ever, descend below the surface but merely walk on the surface film. Such are the springtails, at least two species of which occur on fresh water in Britain. *Hydropodura* (= *Podura*) *aquatica*, the commoner insect, is only about 1·5 mm long, but can

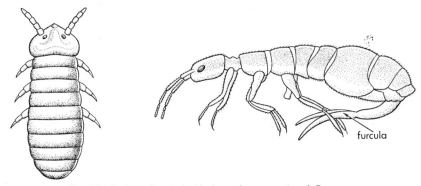

Fig. 64   Springtails: *Left: Hydropodura aquatica*, 1·5 mm.
Right: Isotoma palustris, 1 mm

readily be seen with the naked eye. They are usually found in large numbers, at the margins of still waters, where they might at first easily be mistaken for soot particles on the surface of the water. Underneath their dull bluish-black body is folded a forked tail, which is normally held by a sucker-like appendage between the back pair of legs. When released, the tail hits the water with such force as to project the whole creature several centimetres in the air. Apart from this spectacular means of progression, *Hydropodura* is rather a sluggish creature, crawling laboriously about on its short legs, and occasionally climbing down a plant stem into the water—quite unwetted, for a film of air adheres to its body.

Isotoma palustris, another British species found on the surface of still water, is even small than *Hydropodura*, being about 1 mm in length, but with longer antennae. It is less common.

Little is known about the feeding habits of springtails, but apparently they eat vegetable matter, including pollen, floating on the surface of the water.

Mayflies: Order EPHEMEROPTERA

The mayflies are perhaps by name at least the best known of all the aquatic insects. The brief aerial existence of the adults and the habit of some of the species of emerging from their underwater stages in vast swarms in late May or early June, for their rising and falling mating flight, have been a favourite subject for poets and writers; '. . . gathered into death without a dawn', as Shelley expressed it, and although this is not perhaps literally true, nevertheless their brief adult life is soon spent once its sole purposes of mating and egg-laying are achieved.

The adult insects are easily recognized by the manner in which they hold their four wings erect when at rest, and by their long slender tail filaments, three in some species, two in others.

After a nuptial flight which takes place soon after the female insect emerges from its aquatic stage, the eggs are laid almost immediately, some species merely dropping the eggs into the water, others dipping their abdomens in the water as they settle on it or fly over and washing the eggs off as they are extruded, while still others crawl down into the water and deposit the eggs on submerged objects.

There is much variety in the form and mode of life of mayfly nymphs, illustrating clearly their adaptation to the kind of habitat in which they live. Species living among vegetation in the stiller waters of ponds, lakes or slow streams cling to water-plants, but can swim readily; these include *Cloëon* and *Siphlonurus*. The species of *Ephemera*, found in rivers or lakes with a silted bottom, have flattened, shovel-like fore-legs and upper jaws shaped like daggers for burrowing in the substratum. Agile, streamlined nymphs such as *Baetis* occur in fast streams with vegetation. In faster waters, nymphs with flattened bodies, and adapted to cling limpet-like to the underside of stones, to avoid being swept away, are the characteristic kinds; these include *Ecdyonurus* and *Rhithrogena*. In the silt or moss of small, fast streams are stiff-legged species such as *Ephemerella*. (Plate 34, 1, 5)

The nymphs are mainly active at night, and those that live under stones, in

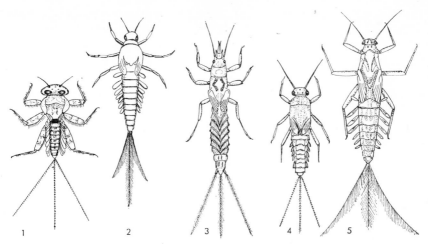

Fig. 65 Mayfly nymphs. 1 *Ecdyonurus*, 12 mm. 2 *Cloëon*,
8–10 mm. 3 *Ephemera*, up to 25 mm. 4 *Ephemerella*, 10 mm.
5 *Baetis*, 15 mm

fast streams, come out of their hiding-places then and are often swept downstream by the force of the current. They become part of the invertebrate drift
which is such an important part of the food of fish, for example the trout.

Mayfly nymphs are readily distinguishable from the aquatic stages of other
insects by their possession of *three* long tail appendages. In those species which
have only two filaments in the adult stages, however, the middle filament of the
nymph degenerates in the final moult.

They all possess tracheal 'gills' on the abdominal segments of the body, but
these vary in form considerably according to the genus, some being tufted and
others, as in *Cloëon*, plate-like. The function of the latter type is largely, if not
mainly, to keep a constant flow of water moving over the body by vibrating
rapidly, respiration taking place through the thin chitinous body covering.

The food of mayfly nymphs consists mainly of vegetable matter, including the
algal coating of stones, although some are believed to be partly carnivorous.
The smaller nymphs complete their metamorphosis in one year or in some cases
less, and there may be two generations in a year. Others, such as *Ephemera*, may
take at least two years.

Transformation usually takes place in the evening either at the surface of the
water, or on a stone or plant out of the water, depending on the species. The
creature which emerges from the nymphal skin, however, is not fully adult, but
a sub-imago with clouded wings and dull in colour—the 'dun' of the angler.
The mayflies are the only insects in which this sub-imaginal stage occurs. After
a short period the final moult takes place, and the skin which veiled the true
splendour of the fully adult insect is shed. The true imago or 'spinner' then

emerges and flies away to reproduce its kind. In some species the male spinners gather in large swarms and indulge in a rhythmic rising and falling flight that is delightful to watch. When a female appears, some of the males detach themselves from the swarm and follow the female; pairing takes place, often at a considerable height in the air.

There are forty-seven British species of mayfly belonging to eight families. They are of importance in the diet of freshwater fish, principally in their aquatic stages, although they are taken even when adult. The importance which the flyfisherman attaches to them is shown by the large number which have been given popular names, a selection of which is given in Appendix II.

Dragonflies: Order ODONATA

Dragonflies must be quite the best known of all insects frequenting fresh water. Their brilliant body colourings, their shining, membraneous wings and the powerful flight of the larger species force their attention on even the most unobservant countrygoer. The larger kinds, too, are found at considerable distances from water, and it is difficult to imagine that they had their origin in some muddy pond or stream.

It will soon become obvious to anyone who takes more than a casual interest in pond-side creatures that there are two distinct groups of dragonflies: the large ones which rest on the pond-side vegetation with their two pairs of dissimilar wings outstretched, and fly strongly and speedily —the so-called hawker and darter dragonflies; and the slender-bodied kind which rest with their two pairs of similarly shaped wings over their backs, and have a fluttering butterfly-like flight—the damselflies. These two groups represent respectively the two suborders of British dragonflies—Anisoptera (Gr. = unequal wings) and Zygoptera (Gr. = yoke-wings).

Both kinds of dragonflies are carnivorous and predacious, devouring large numbers of other insects, including members of their own order. The larger species seem often to have a favourite beat which they patrol for long periods in search of prey, and the vicinity of grazing cattle and horses is also a favourite hunting-ground because of the many insects which are found there. This latter habit has no doubt given rise to the country name of 'horse-stinger' which is sometimes given to dragonflies, although of course these insects have no sting, neither do they bite animals. Damselflies, by reason of their weaker flight, are more restricted in their hunt for food, and confine themselves to the waterside vegetation, darting at small insects at rest on the reeds or other plants. The prey in both groups of insects is dismembered and eaten by the powerful jaws, and the crunching noise made by the larger dragonflies can be distinctly heard some distance away.

An examination of the head of a dragonfly reveals the enormous size of the compound eyes, which occupy almost the entire area of the head. In some species each compound eye contains as many as 8–10,000 facets and, although the image given by these compound eyes can only be a blurred 'mosaic' made up of innumerable points of light and shade, depending upon what is immediately in range of each individual facet, a very good idea of the surroundings and particularly of movements of objects within a range of several yards must be possible. Anyone who has tried on a sunny day to 'net' a large dragonfly resting on a reed will know how difficult it is to get even within range. Nevertheless, in spite of this elaborate arrangement for perceiving their surroundings, dragonflies do not invariably make successful attacks on their intended prey.

The pairing of dragonflies is of particular interest, for it is achieved in a way quite unlike that of any other insect. The male genitals are situated in the last

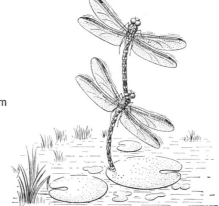

Fig. 66   Damselflies in tandem

but one segment of the body, but by bending its body the male transfers his sperm to special pairing organs on the second segment. The female is then seized and grasped at the back of the head by the male's tail appendages, and in this position the pair fly in tandem to some plant where the female bends back her body until her tail end touches the pairing organs on his second segment and fertilization is achieved.

In some species the insects keep in tandem until egg-laying takes place, the male holding on to a part of a water-plant above the surface, while the female enters the water wholly or partly and deposits her eggs on submerged objects or in incisions she makes in water-plants. Not all species take such care over their egg-laying, however, and the majority merely drop the eggs in the water as they fly over, or dip their abdomens repeatedly in the water to wash off the eggs.

Dragonfly nymphs are common in almost all kinds of still water, some among the water-plants and others on the bottom mud.

The differences between the nymphs of anisopterids and zygopterids are as pronounced as those between the adults. Whereas the former (Plate **30**) are

M

plump-bodied, the latter (Plate 31) are slender, fragile creatures with stick-like abdomens, and are easily distinguished from all other aquatic larvae by having three large, flat, leaf-like tracheal gills at the tail end. One structure all dragonfly nymphs have in common, however, is the so-called 'mask', formed from the fused third pair of jaws or labium, and bearing at its end strong claws. At rest this structure is folded back under the head, but when suitable prey comes within reach, the mask is shot forward and the victim secured by the claws, after which it is slowly eaten.

Both types of nymph are normally sluggish, anisopterids lurking on the bottom of the pond waiting for prey to come within reach, whereas zygopterids are usually clinging to water-plants; in both cases their colouring makes them very inconspicuous against their background.

The leaf-like tail-gills of the damselfly nymphs are traversed by tracheae, and their obvious purpose is respiration; but if by accident or for the purpose of experiment they are lost, the creatures do not seem in any way inconvenienced, and it is concluded that respiration can take place through the thin body-wall by diffusion, although the gills may be necessary in badly oxygenated waters if the nymphs are to survive. When swimming, the gills are waved from side to side so that, should the nymph be deficient in oxygen, a short swim would have the effect of providing a further supply of oxygen.

The anisopterid nymphs have quite a different, and in fact a unique, method of breathing. At the tail end of the body will be seen a number of short spine-like projections. These are round the opening of the anus, and if a nymph is closely watched, it will be seen at times to be taking in water at this point. Inside the body are six double rows of tracheal gills which extract the dissolved oxygen from this water, after which the latter is expelled, often with such a considerable force that the creature is propelled rapidly through the water. This form of jet propulsion is used by the nymphs when they are trying to escape capture. If a nymph is held gently so that the tip of the abdomen is just at the surface of the water, it is sometimes possible to get it to eject the jet into the air, when the distance it travels gives a good indication of the force behind it.

Most of the smaller dragonfly nymphs complete their development in one year, emerging into the air as perfect insects during the summer following the one in which they hatched from the egg. The larger ones, however, may take two years or more, and, as in most aquatic larvae, the exact time taken depends on the abundance of food and perhaps on other conditions in the water.

They moult their skin about twelve times during their development, and around the fifth or sixth moult the wing-buds appear on the back, increasing in size at each subsequent moult. When the time for emergence arrives, usually at night in the case of aeschnids, *Cordulegaster* and *Sympetrum*, but generally in the early morning for other species, the full-grown nymph climbs out of the water, often up a plant stem, and after a short rest the skin is split down the back of the thorax, and the head, thorax and legs of the adult insect are pulled out. After a further rest the abdomen is extricated. The wings at this stage are small and useless, but in an hour or two they have expanded through the circulation of blood into them, and the body itself expands. The insect then flies away in search of food or a mate. Adult dragonflies may live perhaps a month or more, but no British species live through the winter as adults.

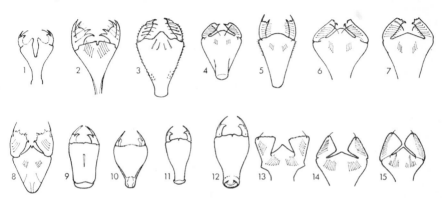

Fig. 67 Masks of dragonfly nymphs: 1 *Agrion virgo*. 2 *Lestes sponsa*. 3 *Platycnemis pennipes*. 4. *Pyrrhosoma nymphula*. 5 *Ischnura elegans*. 6 *Enallagma cyathigerum*. 7 *Coenagrion puella*. 8 *Cordulegaster boltonii*. 9 *Aeshna juncea*. 10 *Aeshna grandis*. 11 *Aeshna cyanea*. 12 *Anax imperator*. 13 *Libellula quadrimaculata*. 14 *Libellula, depressa*. 15 *Sympetrum striolata*

There are forty-three species of dragonflies in Britain, twenty-six of which belong to the sub-order Anisoptera and seventeen to the Zygoptera.

Some of the representative members of the nine families into which they are classified are mentioned below. Identification of nymphs is based mainly on the form of the mask, and in damselflies the gills also.

ZYGOPTERA: Damselflies

AGRIIDAE   The single genus *Agrion* has two species: the demoiselle, *A. virgo*, and the banded agrion, *A. splendens*. The first is more widely distributed and recognizable by the uniformly dark blue colouring of the wings of the male, whereas those of the banded agrion, as the name implies, have broad bands of blue. The females are difficult to distinguish although the wings of *A. splendens* are rather greenish in colour. Both species are on the wing from the end of May to the end of August, *A. virgo* around rapid streams with sandy bottoms, *A. splendens* on stiller waters, including canals and lakes with muddy bottoms. (Plates **31, 32**)

The nymphs of this family are distinguished by having two outer tail gills which are thick and triangular. The mask has a deep cleft in its front edge.

LESTIDAE   *Lestes sponsa* and *L. dryas* are the two British representatives of this family. The latter is rare but *L. sponsa*, sometimes called the green lestes, is common in most parts of the country and recognizable by the colouring which gives it the common name. It flies from June to September, and breeds in still waters. (Plate **32, 5**)

The mask of the nymph is long and narrow at its base. All the gills are thin and rounded at the tip, with heavy pigmentation.

PLATYCNEMIDIDAE   *Platycnemis pennipes*, the only species, is easily recognizable by its broad, white legs, which are dangled by the male during its courtship display. It is a southern species and is seen near running water in June and July.

The nymph is light brown in colour and spotted. The mask is long and triangular, with one short and one long hook on each lateral lobe. The gills are long, lightly spotted and end in a slender, incurved point.

COENAGRIIDAE   This large family has six British genera, comprising twelve species which include some of our commonest damselflies. The large and small red damselflies, *Pyrrhosoma nymphula* and *Ceriagrion tenellum*, are easily recognized by their colour, but the latter species is more restricted to the south and west than the former, which is widespread near still or slow-flowing waters from early May to August. (Plate **32, 3** and **7**) It often seems to become trapped by the sticky leaves of the sundew, *Drosera intermedia*. I was once puzzled by an area of glistening objects near some bog pools, and closer investigation showed them to be the wings of many trapped dragonflies of this species. *Ischnura elegans*, a dark, greenish-black damselfly with blue on the thorax and tip of the abdomen, is on the wing from the end of May until September near still or slow-moving waters. (Plate **32, 6**) *Enallagma cyathigerum* is the commonest blue damselfly and is widely distributed on all kinds of waters, even larger lakes, from May to September. (Plate **32, 2**) *Coenagrion puella*, also blue, although not so common as the last species, is, nevertheless, widely distributed and occurs around still waters from the end of May to mid-August. (Plate **32, 8**) All six species of *Coenagrion* are distinguishable from *Enallagma* by the two black lateral lines on the thorax, whereas the latter has only one.

The nymphs of this large family are not easily differentiated. They all have sharply pointed gills and a mask with a short broad base.

ANISOPTERA: 1. Hawker Dragonflies

GOMPHIDAE   *Gomphus vulgatissimus*, the club-tail dragonfly, is one of the only two large black and yellow British dragonflies, but it is rare. The nymph has a flattened body, well adapted for living partly buried in the mud. It is dark brown in colour and spotted, with a very small head.

CORDULEGASTERIDAE   The golden-ringed dragonfly, *Cordulegaster boltonii*, is the other large black and yellow species, but it has a longer body and, as the common name implies, the yellow is in the form of encircling rings all down the body. It has occurred in most parts of the country, but is not common, and seems to prefer well-aerated waters where there is a silted bottom. It flies actively from June to September. (Plate **33**, 2)

The nymph has a broad head with the prominent eyes near the top so that they project above the mud when the insect is partly buried just below. The mask is triangular and spoon-shaped, and has two broad triangular lobes which have toothed edges.

AESHNIDAE This family includes the commonest of our long-bodied hawker dragonflies, the common aeshna, *Aeshna juncea*. Its body has a mixed pattern of bands, spots and patches of blue, green or yellow which vary according to age and sex. It is widely distributed and commonly seen hawking for its insect prey up and down its chosen beat. It does not appear until July, but it is then on the wing until the end of September or even longer. (Plate **33**, 3)

The nymph lives in weedy ponds or other still waters; its mask is long and straight with one curved hook on each side. The anal appendages are very long.

The brown aeshna, *Aeshna grandis*, is one of our largest species and is common in the south of England and the Midlands. Its yellow wings, blue spots on the top and sides of the abdomen, and often two small blue or yellow spots on the top of the thorax, are distinguishing features. It is on the wing from July to the end of September. (Plate **33**, 1)

The nymph is mottled reddish-brown and olive-green in colour. The head is flat and five-sided when seen from above.

The southern aeshna, *A. cyanea*, is commonest in the south of England, although it does occur in the Midlands and parts of the north. Its colouring is a mixture of brown, blue, green or yellow, according to age and sex, but the males at least are usually distinguishable from the common aeshna by seeming more green on the wing, in contrast to the blue of the commoner species. (Plate **33**, 7) The nymph is similar to the other aeshnids but has a longer, more slender mask.

Our most beautiful dragonfly belongs to this family but is common only in the south of Britain and the Midlands. This is the emperor dragonfly, *Anax imperator*. The adult male has an azure-blue abdomen with a black stripe down the middle. In the female the colour is usually grass-green, but occasional specimens are nearly as blue as the males. They appear in early June and are on the wing until the end of August. (Plate **33**, 5)

The nymph lives in reedy ponds and lakes, and even in brackish water. It is light brown or green, mottled and arched down the middle of the abdomen.

CORDULIIDAE The dragonflies in this family are called emeralds, but except for the downy emerald, *Cordulia aenea*, they are rare. The body is a metallic bronze-green in colour. The downy emerald occurs fairly plentifully in a few localities of southern and eastern England, near ponds and slow streams, and may be recognized by the line of yellow down running the length of the middle of the abdomen.

The nymph lives in the bottom debris and is brown in colour, ornamented with yellowish-green. The mask is triangular and spoon-shaped.

ANISOPTERA: 2. **Darter Dragonflies**

LIBELLULIDAE   Probably the commonest member of this family is the four-spotted libellula, *Libellula quadrimaculata*, which is easily identified by the black or reddish-brown markings in the middle of the front edge of each wing. It is well distributed and the population is often increased by swarms of immigrants from the Continent. It is on the wing from mid-May to the end of August, and although it frequents still waters to lay its eggs, the insect is often seen away from water. (Plate **33, 6**)

The nymph is brown above and yellowish-white below, and rather hairy. The mask is triangular and spoon-shaped, and has broad, triangular side lobes. It lives in the detritus at the bottom of the water.

The broad-bodied libellula, *Libellula depressa*, is well named and can hardly be mistaken for any other species. In adult males the abdomen has a beautiful powder-blue colour with yellow spots along the sides. In females the body colour is reddish-brown. The dragonfly is common south of a line joining the rivers Mersey and Humber, although immigrants sometimes occur further north. It flies from mid-May to early August and prefers still waters even if they are brackish. (Plate **33, 4**)

The nymph is very flat and hairy. The mask projects slightly beyond the head when the insect is at rest.

The common sympetrum, *Sympetrum striolatum*, is the commonest reddish-brown darter dragonfly of late summer. It is widely distributed and large swarms migrate from the Continent to south-east England. (Plate **33, 8**)

The nymph is flattened and spiny, and has a triangular, spoon-shaped mask with broad, triangular lobes.

Stoneflies: Order PLECOPTERA

Stoneflies are found mostly near rapid, clear streams or the wave-washed shores of lakes, which alone provide the well-oxygenated water and type of substratum required by their nymphs. In one or two species of, for example, the family Nemouridae, the nymphs occur in still or slow-flowing waters. The adult insects live only a few days and do not fly readily, so that they are rarely found far from the waters in which they developed and in which the eggs will be laid.

The female flies lay their eggs in May or June, compacting them with a sticky substance into a mass attached to the underside of the tip of the abdomen. The larger stoneflies deposit the eggs into the water by running over the surface, when the mass breaks up and the eggs sink. The smaller species fly over the water, dipping the abdomen so that the egg-mass disintegrates. The nymphs of the smaller species hatch out in about three weeks, but those of the larger species may take as long as three months. Some may be found underneath stones and boulders, others in the mossy covering on the top of large stones, and still others live on the sand and gravel at the bottom of the stream. They may be recognized by their two long tail appendages and by their rather sluggish creeping movement when they are disturbed, which has earned for them the angler's name of 'creeper'. The nymphs range in size from about 5 mm to 25 mm.

Some species are exclusively vegetarian, eating algae and leaves of higher plants, but others eat also small aquatic creatures ranging from Protozoa to insect larvae. The nymphal stages in the larger stoneflies such as *Perla* and *Dinocras* may last three years, but the smaller species complete their metamorphosis in one year. There is no pupal stage, the body of the nymph resembling that of the adult insect, except of course in size and in the absence of wings throughout its development. Breathing is in most species carried out by means of tracheal gills situated along the body in bunches, and if a nymph is observed closely, it will be seen to raise and lower its body rhythmically, presumably to keep a constant circulation of water, and with it dissolved air, passing over the gills. The smaller species breathe only through their cuticle.

After one of the later moults wing-buds appear on the thorax. Shortly before the time for emergence arrives the almost complete wings, crumpled up inside, make the wing-buds quite dark. If a nymph is taken in this stage its transformation into the adult may be watched. Crawling out of the water, it climbs up some vertical support to which it fixes itself firmly with its claws; the skin splits down the back of the head and thorax, and the body of the adult insect is drawn out laboriously, legs and thorax first, then the wings, and finally the abdomen and tail appendages. The wings are soon expanded and, when dry, are folded obliquely along the back. After an hour or so the skin, which at first was soft and light in colour, hardens and darkens. In early summer hundreds of empty nymphal cases of stoneflies may be seen attached to tree trunks and stones bordering suitable waters. (Plate 35)

Adult stoneflies are brown in colour, have long antennae, and retain the two long tail filaments which were such a characteristic feature of the nymph. The females possess four large, heavily veined wings, which in the case of the larger species of Perlidae may measure nearly 5 cm across, but these creatures fly with reluctance. The males of some species, however, have such small wings that they are incapable of flying.

The adults live for only a few days. The larger species do not feed although they drink water. Smaller species feed on lichen and algae which they scrape off trees and fences.

Seven families of stonefly containing thirty-four species are recognized in Britain:

TAENIOPTERYGIDAE   The angler's February Red is the female of *Taeniopteryx nebulosa*, the nymph of which is about 10 mm to 12 mm long and found in muddy rivers, sometimes among the moss. *Brachyptera risi* is a commoner species around small stony streams; the nymphs are found under the stones or in moss.

NEMOURIDAE   These smaller stoneflies occur in slow-flowing or even still waters. They emerge early in the year and are the Early Browns of the angler. *Nemoura cinerea* is one of the commonest and widest in distribution, frequently

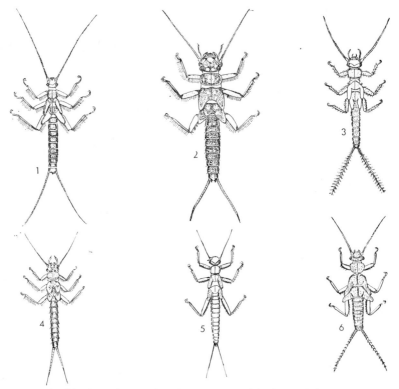

Fig. 68   Stonefly nymphs: 1 *Brachyptera risi*, 10 mm. 2 *Isoperla grammatica*, 15 mm. 3 *Leuctra geniculata*, 10 mm. 4 *Chloroperla torrentium*, 9 mm. 5 *Capnia bifrons*, 9 mm. 6 *Nemoura cinerea*, 5–9 mm

being found in ponds, among the vegetation, during its nymphal stage. It is pale in colour and from 5 mm to 9 mm long. *Nemurella pictei* is common around weedy still or slow-flowing waters.

LEUCTRIDAE   The six species of *Leuctra* range in size from 4 mm to 11 mm and are found in the nymphal stage under stones in rivers and streams. The adults in general are called needle flies. *Leuctra geniculata* is the Willow Fly of anglers.

CAPNIIDAE   The three species of *Capnia* are fairly uncommon, but *C. atra* is sometimes abundant around stony lake shores in Scotland. The nymph is about 6 mm long.

PERLODIDAE   The nymphs of *Perlodes microcephala*, about 12 mm long, are common in stony rivers and streams, as are those of the related *Diura bicaudata*, which is slightly smaller, and found at altitudes of over 300 metres. *Isoperla grammatica*, the Yellow Sally of anglers, is abundant on stony rivers and streams and, in the north, on stony lake shores.

PERLIDAE   The two British species, *Perla bipunctata* and *Dinocras cephalotes*, are The Stonefly of anglers, although probably any large stonefly is so-called, for example *Perlodes* and *Diura*. The nymphs of all of them are known as 'creepers'. The adults are seen in May and June, *P. bipunctata* near rivers and streams with loose stones, whereas *Dinocras cephalotes* prefers those with a more stable and moss-covered bottom. The nymphs are from 14 mm to 33 mm long when full grown. (Plate 35, 1)

CHLOROPERLIDAE   Two species of *Chloroperla*, *C. torrentium* and *C. tripunctata*, are often abundant on rivers and streams with stony bottoms; the adults may be seen until July. The nymphs are from 7 mm to 9 mm long.

The Bugs: Order HEMIPTERA

The order Hemiptera is a large one and only a few of its members are aquatic. Some of these live almost exclusively on the surface film of the water, rarely descending below, whereas others spend their whole life in the water except for occasional flights to other stretches of water. Bugs have three stages in their development: egg, larva and adult. There are usually five larval instars identifiable by the stage of development of the thorax; the wings, where present, become functional only at the final moult.

The most characteristic features of a typical bug are the mouth-parts, which are modified into the form of an elongated piercing and sucking beak or *rostrum*, which in some members of the order is tucked under the head and in others protrudes forward. Bugs are able to take only liquid food, and almost all the aquatic species feed on the juices of other creatures.

All aquatic bugs belong to the sub-order Heteroptera, distinguished by differences between the fore and hind wings, the latter being transparent whereas the former are partly horny and partly transparent; the horny portion forms a protective covering for the hind wings when the insect is not flying. Some water bugs are wingless, however, even in the adult stage. The males of some families have stridulating organs on the thorax with which they make sounds to attract the females during courtship.

HYDROMETRIDAE   Of the bugs living on the surface film of the water and among the marginal vegetation, perhaps the most amazing are the water measurers or water gnats, *Hydrometra*, of which two species occur in Britain. *H. stagnorum*, the commoner species, is about 12 mm long, with a body so incredibly thin that one cannot help wondering how all its internal organs can be accommodated. Its head is greatly elongated, and its eyes are placed well back near the body. When walking slowly over the surface, as if pacing out the distance, the body is held well above the water and the insect apparently never descends below. In this insect, as in other bugs which walk on the surface film, the legs and undersides of the body are covered with a pile of fine hairs which prevent them from being wetted. (Plate **36,** 4) *H. gracilenta,* about 9 mm long and lighter in colour than the common species, is known only from the Norfolk Broads. Both insects are wingless.

Hydrometra feeds on small creatures found either on the water or just beneath the surface film. The female lays her eggs individually on plants and other objects near the water's edge. (Plate **36,** 4)

VELIIDAE   The water crickets, *Velia,* are commoner and more easily seen than the water measurers, for they seem to spend more time actually on the water surface. Large numbers are often found in slow-running streams, near the banks or in little backwaters even during the winter. *V. caprai* is only about 7 mm long, with a stouter body than *Hydrometra,* blackish-brown in colour but with beautiful orange-red markings down the back in the adult. The undersides of both the adults and immature nymphs are orange. Both winged and wingless individuals occur. They feed on small insects that have fallen on to the water surface. The eggs are laid on waterside vegetation. (Plate **36,** 1)

V. saulii seems to occur on more open waters but is less widely distributed.

The three species of *Microvelia* are smaller insects which are usually wingless, but occasionally winged specimens of either sex are found. Their eggs are deposited on floating plants in long rows. *M. reticulata* is the commonest species and is often abundant in thick waterside vegetation.

MESOVELIIDAE   *Mesovelia furcata* is the only British species in this family. It is a small bug, about 3 mm long, with a greenish tinge, and lives mainly on floating leaves in ponds in the south.

HEBRIDAE   *Naeogeus ruficeps,* only about 1·5 mm long, is common among *Sphagnum* moss at the edge of bog pools or tarns. *N. pusillus,* slightly larger, is known only from the south of England.

GERRIDAE   The pond skaters or water striders, *Gerris,* are larger than any of the foregoing species. (Plate **36,** 3) *G. najas,* one of the commonest of the ten species, is about 15 mm long, but looks larger on account of the wide spread of its long legs. Pond skaters are very common insects on ponds, ditches and lakes, sliding along rapidly over the surface, the depressions made by their legs on the surface film throwing clusters of circular shadows on the bottom on a sunny day. Some are wingless, some have very short non-functional wings, and others have fully developed wings. The eggs are attached to submerged plants in little groups and covered with a jelly-like substance. The pond skaters feed largely on dead or dying insects that fall on the water.

In the species of *Gerris*, and also in those of *Velia* and *Microvelia*, the claws are not, as is usual in insects, at the tip of the legs, but a short distance down. A pad of bristles occupies the apex and this prevents the tip of the leg from breaking through the surface film of the water, a modification of great value to insects that spend most of their time on the surface, instead of among vegetation as do other water bugs.

NEPIDAE Turning from the surface dwellers to those living actually in the water, there are the water scorpions, two species of which, quite dissimilar in appearance, occur in Britain. *Nepa cinerea* is a common insect sometimes attaining a length of 30 mm or more. It is quite flat and dark brown in colour; as it lies on the bottom, at the edge of a pond, it looks like a dead leaf and when handled frequently feigns death. At its rear end is a long tube formed of two half tubes that in life are always joined, but that often separate in set specimens. This appendage, which the uninitiated mistake for a sting, is nothing more than a breathing-tube that the insect pushes up above the surface, and thus brings a supply of air to the spiracles. It also raises its wing-cases slightly to admit air to two shallow grooves, roofed with long, water-repellent hairs, on the abdomen, into which the spiracles open. Pairs of organs on the third, fourth and fifth abdominal segments are sensitive to pressure and are believed to help the insect to orientate its body in relation to depth of water and deviation from the horizontal position. The two front legs are modified into grasping organs, by means of which its prey is captured and held while the juices are sucked out by its beak.

If the dark wing-cases are raised, the dorsal surface of the abdomen is seen to be brick red in colour. In most specimens the wing muscles are insufficiently developed to permit flight. (Plates **36**, 2; **37**, **7**)

The eggs are laid individually in spring, in masses of plant material, usually just below the water surface, each egg bearing a number of thread-like tubes varying from seven to nine at its free end. These keep the eggs supplied with oxygen during their development, but diffusion of the gas from the tissues of the plant stems, in which the eggs are embedded, probably also helps. (Plate **39**, 1) The nymphs are easily distinguishable from the adults, even if not by their size, then by the shortness of the breathing-tube.

The rarer water stick insect, *Ranatra linearis L.*, is about 65 mm in length and the body, instead of being flat like *Nepa*, is round and stick-like. Like its relative, it has a breathing-tube and its habits are very similar. The eggs, which have only two appendages, are placed in rows in incisions in plant stems. This insect is found mainly in weedy ponds and can fly. (Plates **37**, 2; **39**, **4**)

NAUCORIDAE The saucer-bug, *Ilyocoris cimicoides* (= *Naucoris cimicoides*) (Plate 37), has an oval body, about 15 mm in length. In the south of England it is fairly common in the muddy margins of weedy ponds. The eggs are inserted in rows in the stems of submerged plants about May. The bug is carnivorous and can give an unwary collector a painful 'bite'. Although it has fully-developed wings, the flight muscles are so atrophied that the insect cannot fly. An air bubble covers the underside of the body and the upper surface of the abdomen.

APHELOCHEIRIDAE This family has one British species which is remarkable in being adapted to live constantly under water without having to come to the surface to renew its air supply. *Aphelocheirus montandoni*, an insect about 10 mm long, lives mainly in fast rivers with a stony or gravelly bottom. (Plate 38) The ventral, and part of the dorsal surfaces of the body are covered with a pile of very fine hairs, bent over at their tips, and so closely set (over two million hairs to the square millimetre) that the film of air held by them is not lost when the pressure inside the film decreases through the withdrawal of gas by the insect's spiracles. More oxygen, therefore, diffuses into the film constantly from the surrounding water—an example of plastron respiration described on page 161. Special rosette-shaped spiracles with branching tubes take the air into the trachea. The pair of spiracles on the second abdominal segment open into collapsible air sacs. Connected to these spiracles are sense organs sensitive to pressure changes. It is believed that these organs enable the bug to orientate itself in the water so as to avoid regions of low oxygen pressure which would be detrimental to the efficiency of its plastron respiration. The bug sometimes buries itself in the substratum, but is quite capable of swimming. It feeds mainly on insect larvae, especially bloodworms and mayfly nymphs. Egg-laying takes place in late spring and summer, the eggs being attached singly to stones. The nymphs do not have a plastron, or even open spiracles, and breathe solely by diffusion through their skin.

PLEIDAE *Plea atomaria* (*P. leachi*) is the only British member of this family and is about 3 mm long. It is found in large numbers among submerged vegetation in ponds, canals and slow rivers, and feeds on cladocerans. The males have a stridulating organ on the thorax that is used during courtship.

NOTONECTIDAE Perhaps the best known of all the water-bugs are the water-boatmen, *Notonecta*, of which four species are found in Britain. They are about 15 mm long. The commonest species is *N. glauca*, which is widely distributed and abundant. Few who have stood by a pond can have failed to see these creatures resting upside down at the surface, the tip of the abdomen and the fore and middle legs just touching the underside of the surface film. They also swim upside down and the body, keeled along the back, with long oar-like hind legs outstretched, bears a striking resemblance to a small boat and amply justifies the popular name. In an age devoted to speed, the streamlining of this 'boat', the frictionless 'skin' and the wonderfully efficient swimming legs fringed with hair to increase their stroke can perhaps be better appreciated than ever before. (Plates **37**, 1, 3; **38**)

Their bodies are very buoyant, no doubt because of the bubble of air held against the ventral surface by four longitudinal rows of hairs which, by leaning together in pairs, form two passages, one on each side of the body. It is only by swimming actively, or holding on to submerged objects, that the insects can remain below. The air bubble acts as a physical gill (see p. 161), but from time to time the insects must rise to the surface to renew it.

Water-boatmen feed on any living creature they can attack, and do not hesitate to tackle creatures much larger than themselves. Fish are not immune to their attacks, and their piercing beak and toxic saliva can inflict quite a painful prick even on a human finger when they are being handled carelessly.

The females lay their cigar-shaped eggs (about sixty in all, over a period of weeks) in early spring. They are laid individually in slits made in water-plants (Plate **39**, 3), with the exception of *N. maculata*, which glues its eggs to stones or other solid supports and lays them, as a rule, in autumn.

Eggs which were laid in a tank that the writer had under observation took two months to hatch, but no doubt the time varies considerably with the weather conditions. The young larvae were at first white with red eyes, and by no means as buoyant as adults, merely resting on their backs at the bottom of the tank. They have no wings at first, but these develop gradually.

Water-boatmen fly readily from pond to pond, taking very little time to prepare themselves for flight once they have climbed out of the water.

Notonecta obliqua, with darker wing-cases than *N. glauca*, is common and widely distributed. *N. maculata* has mottled wing-cases with an orange metanotum and is fairly common in the south of England. *N. viridis* is mainly a brackish water species.

CORIXIDAE The lesser water-boatmen are a large family with thirty-three British species, similar in appearance to the true water-boatmen, but smaller, with flatter backs and a blunter tail end. They range in size from 3 mm to 13 mm. A further distinguishing feature is that they do not swim on their backs. Representatives of this family are found in all kinds of fresh water and even brackish water, each species, however, keeping to a well-defined habitat. It has been shown that in some lakes, as organic material accumulates on the bottom, there is a succession of species of corixids. Thus, on exposed shores with little detritus, the very small *Micronecta poweri* is the only species found. With more cover *Corixa striata* occurs also and with further organic matter *M. poweri* disappears and *C. distincta* and *C. fossarum* appear. They spend more time at the bottom of the water than do water-boatmen, and have to swim rather than float to the surface to renew their air supply, which is stored on a close pile of unwettable hairs on the abdomen and thorax, first obvious at the third instar. Corixids will often be seen rubbing their hind legs against their backs to aid the diffusion of gas into the air bubble. (Plates **37**, 4; **38**)

Unlike most bugs, corixids rarely use their beaks as piercing organs, but instead suck up small particles of organic matter from the bottom of the water as food.

The eggs are roundish in shape, and are not embedded in plants, but attached singly to the submerged stems or leaves of plants or algae. (Plate **39**, 5)

The adult males make a shrill chirping noise at mating time, and *Micronecta poweri* has earned the common name 'water-singer' for its noisy courtship song. The longitudinal rows of pegs on the inner surface of the tarsi of the front legs of the males are used to hold the female during pairing, the distal pegs being caught under the curved ridges of the females' wing-cases.

15: INSECTS, PART TWO

In the last chapter the insects which were described all had one thing in common
—their development from the egg to the adult had no true pupal stage. In this
chapter, which deals with the remaining orders of insects with aquatic members,
we shall be concerned with those in which a complete metamorphosis—egg,
larva, pupa and adult—takes place.

Alder-flies and Sponge-flies: Order NEUROPTERA

SIALIDAE   The classification of the alder-flies has undergone several changes,
but they are now regarded as belonging to the sub-order Megaloptera.

In Britain only two species are aquatic, and those only to the extent of passing
their larval stages in or near the water.

The commonest is the alder-fly, *Sialis lutaria,* the adults of which are familiar
creatures near ponds, lakes and streams in May and June. They are about 25 mm
in length, with long antennae and dull, heavily-veined wings which when the
insects are at rest are folded over the back and ridged like the roof of a house.
These insects fly reluctantly, but when they do take wing they make straight for
their objective without indulging in any aerobatics. They are more likely to be
seen resting or crawling about on the waterside vegetation or on stones near the
water's edge. (Plate **40**)

The females lay their eggs—anything up to 2,000 in number—in neat little
clusters on the leaves or stems of plants near the water or on stones. Each egg
is cigar-shaped and is firmly fixed at one end, the free end bearing a little pedicle
or stalk. They hatch in about a fortnight, and the larvae fall from the egg cluster
to make their own precarious way to the water. (Plate **40, 3**) The larval stage
usually lasts nearly two years and is spent crawling in the bottom mud. When
full-grown, the larva is about 25 mm long, cylindrical in shape and brown in
colour. The thorax is large and clearly divided into three segments, which carry
the three pairs of legs. On the abdominal segments are seven pairs of jointed
structures which at first glance might be mistaken for legs. These are tracheal
gills, and in life the larva carries these curved upwards and backwards. The last
segment of the abdomen tapers away to a point and is believed also to act as a
tracheal gill. Frequently when kept in a small quantity of water the larva will be
seen to undulate its body in a rhythmic manner, no doubt to bring fresh currents
of water over its tracheal gills.

The larva is carnivorous throughout its life, attacking and seizing with its powerful mandibles any smaller creatures it encounters. It will readily attack a stick or camel-hair brush placed near its head, curving its body and retreating in a characteristic jerky manner if 'pursued'.

When fully developed, it crawls out of the water into the mud or vegetable debris near by, and there makes an oval cell in which to pupate, the pupa emerging in about three weeks, when the transformation to the adult fly takes place.

In running water occurs the larva of another species, *S. fuliginosa*, darker in colour and more local in distribution. It has been observed that, in pairing, the males of this species produce a capsule containing sperms, which the female subsequently devours in part. It is possible that careful observation would reveal that the males of the commoner species also produce *spermatophores*, as these capsules are called.

Sub-order PLANIPENNIA

This sub-order includes the lacewings, the ant-lion flies and the two families described below. Only three species are aquatic and one semi-aquatic. They are not rare insects, but are usually overlooked.

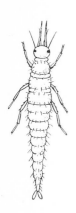
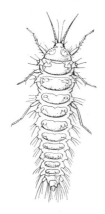

Fig. 69   Left: larva of *Osmylus fulvicephalus*, 15 mm. Right: larva of *Sisyra fuscata*, 5–6 mm

OSMYLIDAE   The larvae of *Osmylus fulvicephalus* live a semi-aquatic existence in the wet moss at the edge of streams. They are about 15 mm long with a rather plump body bearing numerous bristles and tapered at both ends. At the tail end are two short appendages covered with recurved hooks. These larvae feed by sucking the blood of small larvae, such as bloodworms, *Chironomus*, which they find in the moss.

About the end of April they spin yellowish cocoons of silk among the moss and there pupate. In due course the pupa emerges from the cocoon and, resting on a solid support, performs its final moult into the adult fly. The adults are quite large insects with a wing expanse of 50 mm, rather like lacewing flies in

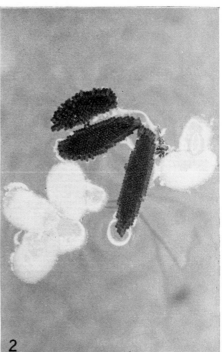

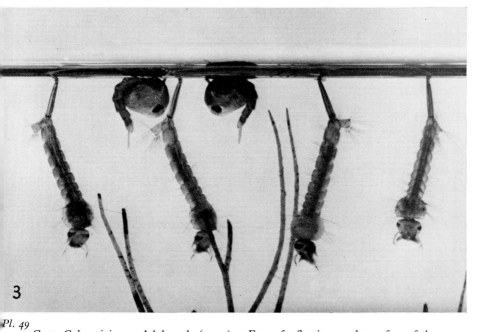

Pl. 49

Gnat, *Culex pipiens*. 1 Adult male ($\times$ 10). 2 Egg-rafts floating on the surface of the water among duckweed, *Lemna minor* ($\times$ 7). 3 Larvae and pupae at the surface of the water ($\times$ 5).

Pl. 50

DIPTERA

1 Rat-tailed maggot, larva of the drone-fly, *Eristalis* (× 2). 2 Pupa of *Eristalis* (× 2).
3 Adult drone-fly, *Eristalis* (× 2). 4 Larva of *Stratiomys* at the surface (× 2). 5 Pupa of
Simulium on a submerged blade of grass (× 7). 6 Larva of *Ptychoptera* (× 1½)

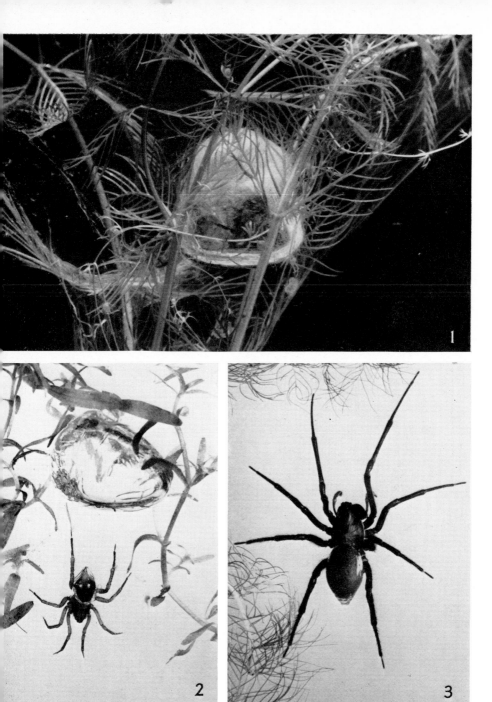

Pl. 51
Water spider, *Argyroneta aquatica*. 1 Bell (× 3). 2 Female spider with partly completed bell (× 2). 3 Male spider (× 2)

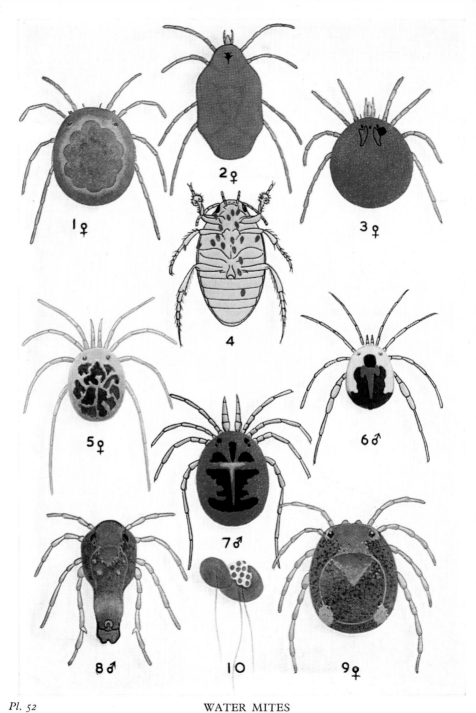

Pl. 52

WATER MITES

1 *Hydrodroma despiciens.* 2 *Limnochares aquatica.* 3 *Hydrachna globosa.* 4 Great diving beetle, *Dytiscus marginalis,* with many larvae of *H. globosa* attached. 5 *Hygrobates longipalpis.* 6 *Neumania spinipes.* 7 *Limnesia fulgida.* 8 and 9 *Arrhenurus buccinator,* male and female. 10 Eggs of *Hygrobates longipalpis* on duckweed, *Lemna minor.* (Mites approx. 2 mm in length)

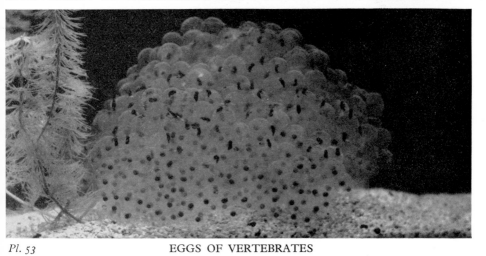

Pl. 53   EGGS OF VERTEBRATES

Top: Eyed ova of trout, *Salmo trutta* ($\times$ 2). *Middle:* Egg-rope of common toad, *Bufo bufo* ($\times$ 1). *Bottom:* Spawn of common frog, *Rana temporaria* ($\times$ ½)

Pl. 54

1 Tadpoles of common frog, *Rana temporaria*, with legs ($\times$ 1½). 2 Larva of smooth newt, *Triturus vulgaris* ($\times$ 3 approx.). 3 Male and female smooth newts

Pl. 55

COLLECTING EQUIPMENT (1)

A. Square plankton-net. B. Weed drag. C. Bottle-clip. D. Weed-cutting knife. E. Pondside trough. F. Forceps. G. Concentrating bottle. H. Pipette. I. Angler's bait-tin. J. Collecting bottle. K. Magnifying lens. L. Specimen tube in corrugated paper. The dish is also a useful accessory for examining the catch

Pl. 56
    Above: Smooth newt, *Triturus vulgaris* at the water surface. *Below:* Three-spined sticklebacks, *Gasterosteus aculeatus*. (Both slightly enlarged)

Pl. 57 FROGS AND TOADS

1 Common frog, *Rana temporaria*. 2 Edible frog, *Rana esculenta*. 3 Marsh frog, *Rana ridibunda*. 4 Natterjack toad, *Bufo calamita*. 5 Common toad, *Bufo bufo*. (Approximately natural size)

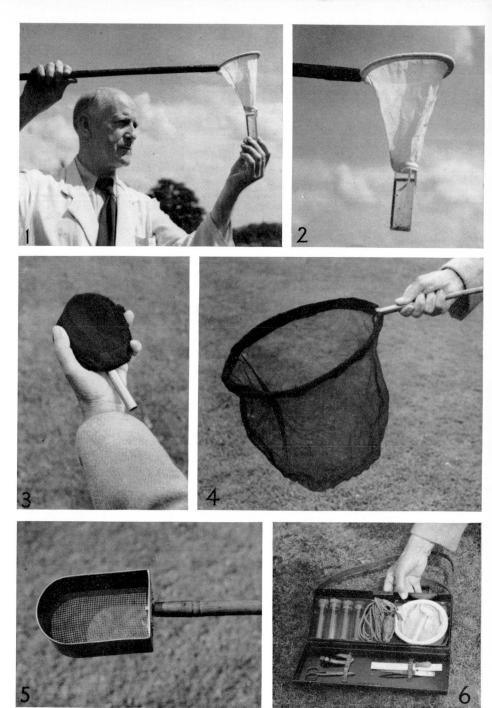

Pl. 58 COLLECTING EQUIPMENT (2)

1 and 2 Small plankton-net. 3 Butterfly net, folded. 4 The same open and ready for use.
5 Shell scoop. 6 A convenient collecting set with net, drag, scissors, forceps, weed knife,
pipette, ruler and tubes

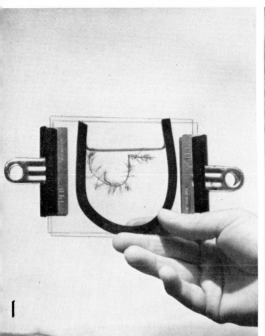

Pl. 59

1 Improvised pondside trough. 2 A coarse muslin strainer fitted into the top of a net. 3 Stone-turning at the edge of a mountain stream. 4 Floating cage for capturing insects emerging from aquatic nymphal or pupal stages

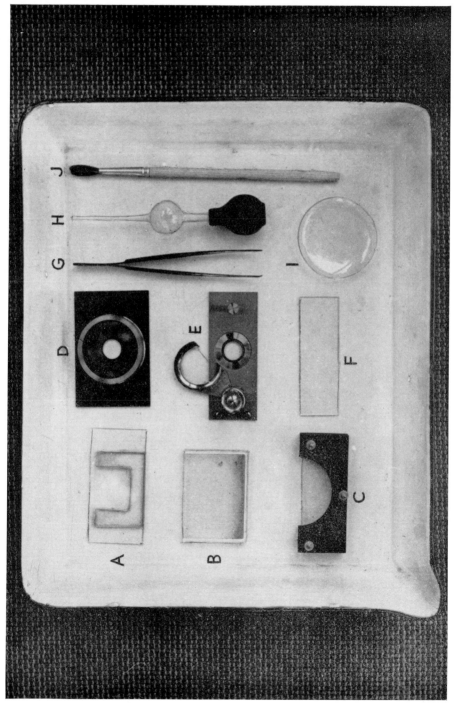

EQUIPMENT FOR MICROSCOPICAL EXAMINATION OF AQUATIC LIFE

A. and B. Stage-troughs. C. Botterill's trough, using two ordinary microscope slips separated by half

Pl. 60

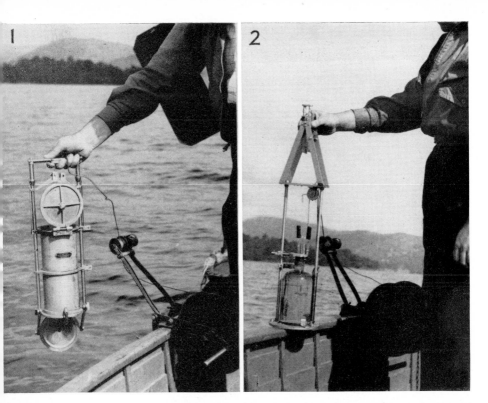

Pl. 61 EQUIPMENT FOR LAKE INVESTIGATIONS

1 Water-bottle which can be closed at any required depth. 2 Water-bottle used for obtaining bacteriological samples from any required depth. 3 Thermistor with galvanometer and battery, for rapid routine measurements of water temperature

Pl. 62

Above: Jenkin surface mud-sampler for taking an undisturbed section of the mud surface of lakes for chemical and biological examination; in the background is a large tow-net for collecting plankton samples. *Below:* Mackereth oxygen probe for the rapid and continuous determination of oxygen concentrations in the water

Pl. 63
    Above: Primary filters at a waterworks. *Below:* Secondary filters at a waterworks. The one in the foreground has been emptied for the purpose of cleaning the sand

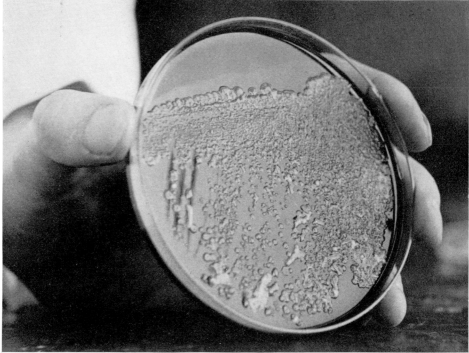

Pl. 64
Above: The interior of an empty primary filter at a waterworks. The zooglea layer occurs at the bottom of the large rectangular compartments. *Below:* A petri-dish culture of the bacterium *Escherichia* (= *Bacterium*) *coli*. This organism, which is a normal inhabitant of the human intestinal tract, is used to assess contamination of water

general appearance, but their wings are veined with dark brown instead of green and have blackish spots on them.

In *Osmylus*, as in *Sialis fuliginosa*, the males produce a spermatophore which is partially devoured by the female after pairing has taken place. It is believed that the males attract the females to them by means of scent glands on the abdomen.

The eggs are laid in clusters on leaves near the water, but, unlike those of the alder-flies, they are placed on their sides.

SISYRIDAE The larvae of the sponge-flies, of which three species occur in Britain, are widely distributed throughout the country, but are rarely encountered by the casual observer, since they live on the surface or in the cavities of fresh-water sponges, feeding by sucking up the juices of these creatures. When full-grown they are about 7 mm long, and have plump bodies tapering to the tail and bearing tufts of hairs. On the abdominal segments are seven pairs of tracheal gills which are constantly vibrated.

In spite of the unpromising diet on which the larvae exist, they grow rapidly; some are ready for pupation when a few weeks old. They then crawl out of the water on to a tree trunk or other solid object, where they spin a cocoon, the adult flies emerging in about a fortnight to produce a second brood in late summer. Many larvae, however, do not become full-grown until late summer, and these do not emerge as adults until next spring, at which time also appear those from the second brood.

The adult sponge-flies resemble lace-wing flies, but are only about 6 mm in length with brownish-coloured wings. The commonest species is *Sisyra fuscata*.

The Beetles: Order COLEOPTERA

There are more species of beetles in the world than of any other creature, and they have adopted the most diverse types of environment. It is hardly surprising, therefore, that some of them have invaded the water and, although few British beetles spend their whole life there, all members of at least nine families, and a few scattered representatives of other families, are aquatic at most stages of their development.

The most characteristic features of a typical beetle are the fore pair of wings which have become hard and horny and quite useless for flying; they now serve merely as protective sheaths for the hind pair. These *elytra*, as they are called, meet in a straight line down the back and cover the soft and flexible upper surface of the body. The name of the order is derived from this feature (Gr. *koleos*, a sheath; *pteron*, a wing).

The true wings are membraneous and usually quite long, so that when not in use they have to be folded and creased to pack them away under the elytra.

A comparison between a typical land beetle and one of the true water beetles, such as *Dytiscus marginalis*, will soon show how well adapted the latter is for an

N

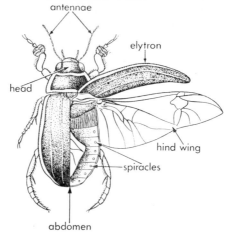

Fig. 70 Structure of a water beetle, *Dytiscus*

aquatic existence. The body is beautifully streamlined for rapid progress through the water, and the legs, particularly the hind pair, are well clothed with hair, which greatly increases their effective surface when they are used for swimming.

In some of the families, too, the basal joints of the hind legs are much expanded and modified to provide a large area for the attachment of muscles, thus enabling powerful and vigorous swimming strokes to be made, although preventing motion of the legs in any direction other than the horizontal, so that when a water beetle tries to walk on land it can only do so in a very clumsy manner.

Except for some species of the family Elminthidae (p. 192), no adult British water beetle can obtain all its oxygen supply direct from the water, and the majority therefore depend mainly on atmospheric air, for which they frequently rise to the surface. Spiracles, or air-holes, are situated along the body under the wing-cases, but in some species, such as *Dytiscus*, the two pairs of spiracles near the tip of the abdomen are considerably larger than those on the rest of the body.

When the beetle rises to the surface, this area of the body is clear of the water and the spiracles are able to take in air. At the same time the wing-cases are raised slightly, to renew the air bubble trapped in the pile of fine hairs on the upper surface of the abdomen, which serves as a reservoir and enables the beetle to remain submerged for quite long periods. (See p. 161.)

The larvae of water beetles are also mainly dependent on atmospheric air for breathing and, like the adults, rise to the surface to obtain it. The pupal stage is usually spent out of the water, so that no problem of respiration occurs.

The order Coleoptera is divided into two-sub-orders: Adephaga, sometimes referred to as the carnivorous beetles; and Polyphaga, the so-called omnivorous beetles. The most obvious differences between the two are that the first have

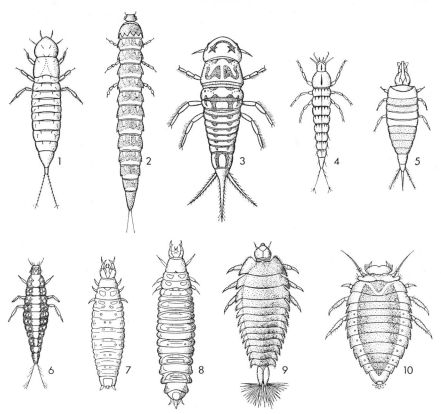

Fig. 71 Larvae of water beetles: 1 *Hydroporus*, 9 mm. 2 *Haliplus*, 6 mm.
3 *Hygrobia*, 17 mm. 4 *Agabus*, 18 mm. 5 *Hyphydrus*, 7 mm. 6 *Ilybius*, 15 mm.
7 *Hydrobius*, 9 mm. 8 *Laccobius*, 3 mm. 9 *Elmis*, 4 mm. 10 *Helodes*, 7 mm

thread-like (filiform) antennae and carnivorous larvae with clearly defined tarsi
(the terminal joints of the legs), usually bearing claws; the Polyphaga, whose
feeding habits of both adults and larvae vary in different species, have diverse
antennae and larvae whose legs end in a single claw.

The families in each sub-order which have aquatic representatives are given
below with brief descriptions of a few typical members.

Sub-order ADEPHAGA (Gr. *ade*, satiety; *phago*, to eat, the implication being
that they are greedy eaters)

HALIPLIDAE   The sixteen species of *Haliplus* are all small beetles, none of
them more than 4 mm in length, and yellowish or reddish in colour. (Plate **44**,
2) They are found mainly in ponds and ditches, although some species occur in
running water and even in brackish water. The adults are not obviously adapted
for life under water, and spend most of their time clambering about the water-

plants, particularly the algae on which they feed. They swim feebly, moving the legs of opposite sides alternately. The basal joint on each pair of the hind legs is expanded into a large plate, and air is collected under this when the beetles rise to the surface. The air passes along a groove into a space beneath the elytra, whence it is taken in by the spiracles along the body.

The white oval eggs are attached to algae and hatch in about ten days. The larvae are very thin creatures about the same length as the adults when full-grown, their bodies tapering to a point at the tail end. They, too, are found among vegetation, and feed by abstracting the contents of the cells of filamentous algae. They renew their air supply at the surface, although the larvae of another genus in this family, *Peltodytes*, are able to extract oxygen from the water by means of tracheal gills.

The pupation period is spent in a cavity in the soil near the water's edge. Instances are on record, too, of the larvae being found hibernating under stones on land during the winter months.

HYGROBIIDAE   The only British species in this family is *Hygrobia* (= *Pelobius*) *hermanni*, which will soon announce its presence in a collecting net by making a squeaking sound, caused by rubbing the roughened surface on the underside of each wing-case against the sharp edge of the abdomen. This accomplishment of both the males and females has earned the insect its common name of screech beetle. (Plate **41, 5**)

It occurs in ponds rich in decaying matter, particularly those frequented by cattle, and is carnivorous. The eggs are laid in spring in rows on the surface of submerged plants, and the larva can easily be recognized by the large head and fore-part of the body, the tapering abdomen and the three hairy tail filaments. It possesses tracheal gills and is not dependent on atmospheric air for breathing, spending most of its time hunting in the mud for the *Tubifex* worms on which it lives. The pupal stage is passed in the soil at the water's edge.

DYTISCIDAE   The members of this family are the water beetles *par excellence*. Their bodies are well suited for rapid progression through the water, and they may be distinguished from other families by their thread-like eleven-jointed antennae and by the strong mandibles with which their animal prey is captured and eaten. A typical example is the great diving beetle, *Dytiscus marginalis* (Plates **41**, 6, 8; **42**), the largest and best known of them all. More has surely been written about this beetle than any other. One German work alone, dealing exclusively with this insect, runs to two volumes and a total of 1,827 pages!

D. marginalis is common and may be found in ponds all over Britain. The adults are about 35 mm long, olive-brown in colour and with a yellow margin round the wing-cases and thorax, from which the trivial name *marginalis* is derived. The sexes can be distinguished firstly by the texture of the wing-cases, which in the male are smooth and glossy, those of the female usually being furrowed along their length; and secondly by the great expansion in the male of the first three segments of the tarsi of the front pair of legs into a circular pad. (Plate 42, 3) This pad bears two large suckers and many small ones, which can all be moistened by a sticky secretion. During mating the sucker-pad is applied to the smooth pro-thorax of the female and the middle legs to the sides of her elytra, and in this way I have seen captive specimens swim together, the male carrying the female, for hours at a time.

The beetles are ferociously carnivorous, attacking any creatures, even if much larger than themselves, and are frequently the cause of the mysterious death of fish kept in garden pools. In summer they occasionally indulge in nocturnal flights, during the course of which they often come to grief through mistaking a greenhouse roof or even a wet road for a new stretch of water.

They swim rapidly with powerful strokes of the fringed hind pair of legs beating in unison, but when resting they must hold on to submerged objects, for they are very buoyant because of the air bubble between the elytra and abdomen, and quickly rise to the surface if not anchored. The rear end of the body is the lightest part, and when the beetle reaches the surface the two large spiracles here are brought into communication with the air. The value of such an arrangement to a damaged insect can be realized. Merely by ceasing to exert itself, it rises to the surface in the correct position for breathing.

The larva when full-grown is about 50 mm long, with an elongated body. At its tail end are two appendages fringed with hairs. The head bears a pair of powerful sickle-shaped mandibles through each of which runs a hollow tube for the injection of a digestive fluid into the prey, and the subsequent sucking up of its dissolved body contents. The larva, like the adult, is lighter than water, particularly at the tail end, and when it is at rest, holding on to a submerged object, the abdomen is held in a characteristic upright position. When the larva rises to the surface, the tail filaments adhere to the surface film and the creature takes in air through the two tail spiracles which are thus brought into communication with the atmosphere.

The larva is, if anything, even more ferocious than the adult, and similarly disposes of any living creatures that it encounters. The unfortunate victim is seized by the mandibles, and not only are its juices sucked out, but a digestive fluid is pumped into its body, the solid contents dissolved and then sucked back, to leave only an empty skin at the end of a meal.

The larval period lasts longer than was formerly thought. Full-grown larvae may be found as early as May, and young ones as late as October, which seems to indicate that the larval period occupies about a year. Pupation takes place in an oval cell in the soil near the water's edge and lasts about three weeks, but the newly-emerged beetle remains in the pupal cell for a further week or two until its body-covering hardens.

The females lay their sausage-shaped eggs about March, individually in slits made in the stems of submerged plants.

In winter the adults may often be found by scooping a net through the mud at the bottom of the pond, and I well remember my elation as a boy in obtaining in this way six of these beetles in one sweep of the net!

When specimens of *Dytiscus* are handled, they will sometimes try to nip one's finger, while a further means of defence is the discharge of an evil-smelling white fluid from the thorax. A yellow fluid smelling of ammonia is also said to be discharged from glands at the rear end of the body. The adult beetles are often infested with larvae of a red water mite, *Hydrarachna globosa,* many specimens of which may be attached to the underside of the body. (Plate 52, 4)

There are five other species of *Dytiscus*, all smaller than *marginalis*, but with similar life-histories.

A relative of *Dytiscus*, which is very common in ponds, is *Acilius sulcatus*, a beetle about 16 mm in length with a flattish brown body and pronounced black markings on the head and thorax. The larva is similar to that of *Dytiscus*, but has an exceptionally long first segment of the thorax, so that it appears to have a long neck. Its eggs are dropped at random in the mud. (Plates 43, 4; 44, 7, 9)

In addition to these species of larger beetles, the family Dytiscidae includes several genera of small species, many of which are common in ponds and other still-water habitats. A few typical representatives of these are illustrated on Plates 41 and 44.

Of the three species of *Laccophilus*, the commonest is probably *L. hyalinus*, of interest because of the stridulating organs on the base of the back legs of both sexes. It is a running-water species about 4 mm long. *L. variegatus*, slightly smaller, is found in drainage dykes, mainly in the south. (Plate 41, 3) *Hyphydrus ovatus* is a widely distributed species, found in still backwaters of streams, silted shores of lakes and similar habitats (Plate 44, 6). The genus *Hydroporus* has thirty-three British species. *H. erythocephalus* (Plate 44, 4) is common in boggy and marshy habitats, including salt-marshes. Some of the nineteen species of *Agabus* are very common in ponds. They have shiny bodies between 6 mm and 10 mm long, and are blackish or reddish in colour. *A. bipustulatus*, one of the largest, is black with red antennae, although reddish forms are sometimes found. It is common everywhere in ponds and still waters generally. *A. uliginosus* is often found in grassy pools with soft, muddy bottoms. (Plate 44, 3) *Platambus maculatus* (Plate 41, 4) about 8 mm long, is a common and easily identified species with its pronounced yellow and black markings. It is usually found among submerged vegetation at the margins of running water. The mud dweller, *Ilybius ater* (Plate 41, 1) is widely distributed in stagnant waters, including peat pools and brackish habitats, and is recognizable by the orange markings on the edges of the elytra. *Rantus notatus* (Plate 44, 1) is one of the six species of its genus and is found in ponds with silted bottoms, in widely separated parts of the country.

GYRINIDAE   The last family in the sub-order Adephaga is that containing the well-known whirligig beetles, *Gyrinus*, those small oval beetles with very shiny black wing-cases which are usually seen, particularly in late summer, in groups on the surface of ponds or slow streams, whirling in intricate patterns when disturbed. Their mazy gyrations, primarily important perhaps in finding food, are also probably of benefit to the species in another way, for no sooner is one beetle aware of the presence of an intruder, than it starts swimming wildly about, and thus, no doubt, warns the others in the vicinity, for they soon do the same. If the danger becomes imminent, as when a net is wielded among them, the beetles will disappear, some diving below, others scattering wildly. (Plate **41, 7**)

They are essentially creatures of that intermediate state between air and water, the surface film, and are well adapted for living there. Their eyes, for instance, are each divided into two parts, the upper of which is believed to see above water and the lower below the surface. Their amazing speed is partly due to the beautifully streamlined body, but mainly to the efficiency of the curiously

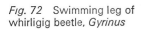
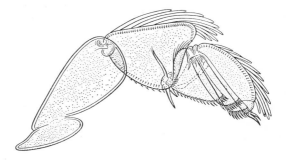

Fig. 72 Swimming leg of whirligig beetle, *Gyrinus*

modified middle and hind pairs of legs, in which each segment is greatly shortened and flattened and fringed with hairs. On the backward swimming stroke these plates present their broad surface to the water, but on the return fold up and offer but little resistance. The front pair of legs is capable of grasping objects.

The antennae of whirligig beetles are very unusual, being club-shaped, with the third joint so large that it resembles a second small antenna joined on to the main one.

Living at the surface, the beetles obtain their air supply readily, but when they dive a silver bubble of air is usually attached to the rear end of the body, and a supply of air is also carried in the space under the wing-cases, making the creatures very buoyant and unable to remain submerged unless holding on to objects below.

Eggs are laid in spring in batches of twenty to thirty, attached to submerged plants. The larva is an elongated creature about 25 mm overall when full-grown, with a pair of hair-like tracheal gills on each segment of the abdomen except the last, which has two pairs. At the extreme tip of the abdomen are two pairs of curved hooks, the purpose of which is not clear, although they may enable the larva to grasp its support when making its cocoon. (Plate **43, 3**)

The larvae are usually found clambering among aquatic plants, sometimes quite deep in the water, and they are believed to take both animal and vegetable food. Towards the end of July, the larva climbs out of the water up a plant stem. and there makes a cocoon of mud or other material in which it pupates. These cocoons must be very numerous, but are rarely seen. The adult beetle emerges

Fig. 73   Cocoon of *Gyrinus* on a plant stem

about the end of August, and at this time of the year whirligigs are particularly abundant. When severe weather comes, they bury themselves in the mud, but they can be seen gyrating on the surface of the water during mild spells in winter.

There are ten species of *Gyrinus* in this country which have been much confused in the past. *G. natator*, *G. minutus*, *G. caspius* and *G. marinus* are species with widespread distributions in Britain. The hairy whirligig, *Orectochilus villosus*, is similar to the species of *Gyrinus*, except that the upper surface of the adult beetle is well clothed with short hairs.

Sub-order POLYPHAGA (Gr. *poly*, many; *phago*, to eat, hence omnivorous)

HYDROPHILIDAE   This family contains both terrestrial and aquatic genera, but among the latter is included one of the best-known water beetles, the great silver beetle, *Hydrophilus piceus*. Its popular name refers to the silvery appearance of the large bubble of air which covers the underside of the thorax and fore-part of the abdomen, and is trapped by two sets of unwettable hairs, one a closely packed pile of minute hairs—the micro-plastron—and the other of long, flexible hairs bent over parallel to the surface—the macro-plastron.

The aquatic members of this family occur in ponds and slow streams, particularly those with plenty of aquatic vegetation. These beetles are nearly all omnivorous in the adult stage, and spend much of their time crawling about among the water-plants and browsing on them. They are poor swimmers and, when they do swim, their legs beat the water alternately and not in unison, as with the dytiscids, and their progress through the water is wavering and unsteady.

The antennae in this family are short, and what at first sight might appear to be antennae are really the unusually long sensitive maxillary palpi, or sensory organs of the mouth-parts, which appear to serve much the same function as antennae do in other insects. The real antennae, which are club-shaped, with the ends clothed in hairs, play an important part in the breathing of the insects. When hydrophilids rise to the surface to replenish their air supply, they do not project the tip of their abdomen above water as do dytiscids, but come up head first and then, inclining their body a little to one side, break the surface film with an antenna. Fringes of hair on the head and antennae form a channel which brings the atmospheric air into communication with the two air reservoirs.

When the beetle is active in the summer months the air film on the 'micro-plastron' of small hairs is inadequate for its needs, and it rises at intervals to renew the supply in the 'macro-plastron' of longer hairs. In winter the micro-plastron may supply most of the beetle's oxygen. At this time of year I have noticed that specimens being studied in an aquarium seem to become water-logged, and on one occasion I took out a beetle that seemed to be dead and put it on one side to set later, only to find that after a few days it had become active again. Other workers have found that the beetles climb out of an aquarium at this time of year in order to groom their plastron with their front and middle legs. Probably, in a natural state, this grooming out of water is necessary at intervals, when the macro-plastron is being used infrequently.

Unfortunately, the popularity of this fine insect as a specimen for the aquarium has led to its disappearance in parts of the country where it used to be found, and it is now confined to a few localities, most of them in southern England. The number of specimens now caught in mercury-vapour light traps in south-east England indicates that our native stock is, on occasions, reinforced by immigrants from the Continent during spring and summer.

It is our bulkiest beetle, for its only rival in regard to size, the stag beetle, *Lucanus cervus*, although a longer insect on account of its antler-like mandibles, has a smaller body.

An average specimen of *Hydrophilus* is about 45 mm long and of a greenish-black colour. The male is readily distinguished by the end joints of the tarsi (feet) of the front legs, which are flattened into the form of triangular plates. Both sexes have a sharp spine on their undersides, which is liable to prick one's hand if the beetles are not handled carefully. (Plates 41, 2; 43, 2)

The female beetle lays her eggs, up to about fifty in number, in a silken container. The spinnerets from which the silk issues forth are situated at the rear end of her body and, clinging to weeds at the surface of the water, she weaves and attaches to a plant a net of silk, which in due course encloses her abdomen. As the work proceeds the eggs are arranged in neat rows hanging from the top of the case, and when they have all been laid the case is sealed and provided at one end with a hollow 'mast', which may be to ensure an adequate air supply for the developing embryos in their impermeable container. The case is usually attached to the aerial shoots or leaves of water-plants, but I have noticed that even when floating loose it seems to be strongly attracted to fixed objects, probably because of some manifestation of the surface tension of the water. (Plate 43, 1) Females may make more than one egg-case in a season.

After about a fortnight, the larvae bite their way out of the egg-case and drop into the water, but return frequently to the surface for air. Unlike their parents, which eat both plant and animal food, the larvae are exclusively carnivorous, and seem to prefer water-snails, especially species of *Lymnaea*, which are seized by the powerful mandibles and carried about until only the shell is left.

The larvae (Plate 43, 5) grow rapidly, and by late summer may be 70 mm long. They leave the water and pupate in a cell which they make in the damp soil near by. The fully-developed beetles emerge about six weeks later.

Hydrochara caraboides, which resembles the great silver beetle in habits and general form, but is only about half the size, is found in similar habitats to its larger relative. *Hydrobius fuscipes* (Plate 44, 5) is smaller still, being only about 12 mm long. The larva of this species has four characteristic teeth on each mandible. *Helochares lividus*, a yellow beetle quite common in weedy ponds, is remarkable in that the female carries the egg container round with her fixed to the abdomen. The egg containers made by the various species of *Philydrus* are round and attached to the leaves of floating plants, the pupation period being passed in the water attached to plants.

ELMINTHIDAE   The beetles in this family are all small, and are found mainly in or near running water.

Both the adults and larvae of *Elmis* live underneath stones and moss in streams and rivers. The legs of the adults bear strong claws with which they can hold on to the surface of the stones and moss, and so avoid being washed away by the current. They obtain their oxygen requirements by plastron respiration (see p. 162). A small bubble of air is usually carried at the end of the abdomen. The larvae are flattish creatures, tapering to the tail end, where there are a number of feathery filaments. When they are being examined under a microscope these filaments are seen to be continually retracted into the body and then extended. They are, no doubt, a form of gill. Both larvae and adults seem to live on vegetable matter such as algae. (Plate 44, 8)

DRYOPIDAE   The various species of *Dryops*, small beetles about 5 mm in length and with hairy bodies, although living in streams during the day, are said to leave the water at night and fly about. The larvae are worm-like and live in rotting debris or damp ground near the water.

CHRYSOMELIDAE   This is a large family, but only a few species are aquatic or semi-aquatic. The species of *Donacia* are beautiful beetles with a brilliant metallic sheen, and are found on waterside vegetation. The larvae, however, live under water, and are remarkable in that they obtain their air supply by tapping the air spaces in the roots of emergent plants. They are fat, inactive, white, grub-like creatures about 10 mm long, and live on or in the roots, piercing the tissues by means of hollow spines borne on the tail end of the body, to reach the air spaces. The spines communicate with the tracheal system of the insect.

The pupae of these beetles are enclosed in oval transparent, brown cocoons, which are also firmly attached to submerged roots, the root forming one side of the cocoon. If one of these cocoons is removed, the surface of the root will be seen to bear a number of fine punctures, indicating that the pupa, too, draws on the plant for its air supply. The pupa changes into the adult inside the pupal case, and eventually makes its way to the surface and out of the water. The eggs are laid on the underside of the leaves of water-plants, the female beetle biting a round hole in the surface of the leaf and depositing the eggs round the edge. (Plate 47, 4, 5)

Macroplea (= *Haemonia*) *mutica* is a small black and yellow beetle, about 5 mm long, which is found in late summer at the roots of aquatic plants such as fennel-leaved pondweed, *Potamogeton pectinatus*. It favours brackish water. The underside of the thorax and abdomen, as well as the long antennae, are covered with a shiny covering of golden hair forming a plastron, which enables the insect to obtain oxygen for respiration without coming to the surface. The larvae resemble those of *Donacia*.

The species of *Galerucella* are not really aquatic beetles, but as they are often found in considerable numbers around the margins of ponds and lakes, the question of their identity often arises on pond-hunting expeditions. *G. grisescens* is orange-coloured and is often found on the leaves of amphibious persicaria out of the water. The yellowish eggs are laid in small clusters of about twenty on the underside of these leaves, and irregular-shaped holes are usually eaten out of the leaves. (Plate 48, 5)

Finally, mention must be made of a genus of small beetles that are not aquatic, but some species of which live near water, and are sometimes seen on the surface film. The *Stenus* beetles, belonging to the family Staphylinidae, are able to propel themselves by secreting a substance from their rear end which lowers the surface tension of the water and they are moved along, presumably using the same principle as toy boats fitted with a piece of camphor. Their legs apparently take no part in their movement for they are held above the surface.

Caddis-flies: Order TRICHOPTERA

Adult caddis-flies are not unlike moths in appearance, and, since they are usually seen on the wing at dusk, they are no doubt often mistaken for these insects. If the wings are examined under a microscope, however, they will be seen to be covered with fine hairs instead of the scales with which the wings of moths and butterflies are clothed. This accounts for the name of the order (Gr. *trichos*, a hair; *pteron*, a wing).

They are mostly of sombre hue, browns and greys predominating, and their long and many-jointed antennae are distinguishing features. The life of the adult flies is brief and is spent near the water from which they came. They eat little or nothing, their jaws being atrophied, although some species can imbibe liquids. Fly-fishermen call some of them sedge flies, and their larvae form a large part of the diet of trout.

There are two main types of caddis larvae. The most familiar one is the so-called *eruciform* (L. *eruca*, a caterpillar; *forma*, shape) type, which makes for itself a protective case from pieces of stick, leaves, stones, sand grains or even snail-shells—sometimes with the snail inside! Each kind of caddis usually has its characteristic type of case, but instances have been known of a larva occupying the empty case of another species. Always there is an inner tube of silk, spun from the mouth, and to this the other material is attached.

This type of larva has a soft, light-coloured abdomen, bearing on the first segment three protuberances which serve to grip the inside of the case. Along the body are tracheal gills with which the creature can extract oxygen from the

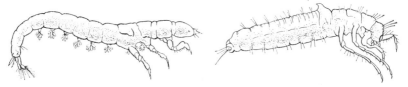

Fig. 74   Caddis larvae : *Left:* campodeiform type, *Hydropsyche*.
Right: eruciform type, *Limnephilus*

water, and often the larvae will be observed undulating their bodies inside the case to keep a current of water circulating over the gills. At the rear end of the body are two blunt claspers which hook into the silken lining of the case so firmly that it is often difficult to remove a larva without damaging it. The insertion of a pin *head* into the rear end of the case will usually be found effective. The fore-part of the body, including the head, is covered with a thick chitinous layer, which gives it a dark appearance in contrast to the light abdomen.

The larva walks about at the bottom of the water, its head and legs protruding from the front of the case, but as soon as danger threatens, the whole body is instantly withdrawn. Even this tactic does not afford protection against fish, which swallow the creature, case and all. One species, however, *Anabolia nervosa*, seems to have solved this problem, and walks freely about on the bottom of the stream in broad daylight, unmolested. Its case has attached to it long pieces of stick, and no doubt these make it too formidable a mouthful even for greedy trout.

The case-building larvae feed almost exclusively on plants, although they do occasionally eat animal matter. As they grow, their cases are extended by adding new material at the front end. The rear end has only a small aperture, just sufficient to allow a current of water to pass through the case.

The larval period occupies about a year, and when the time for pupation arrives, the larvae construct a grating of silk or vegetation over the end of their cases which, although it allows water to pass through, prevents the ingress of undesirable visitors. Sometimes the cases are fastened to submerged stones. The pupation period lasts some weeks at least and may even extend over the winter months in some conditions. When the time for emergence arrives, the pupae bite their way out of the case with their powerful mandibles and either swim to the surface or crawl up the bank out of the water, where the adult fly breaks out of the pupal case and flies away.

Sometimes a caddis case is found with a long filament attached. This indicates the presence of a small parasitic ichneumon fly, *Agriotypus armatus*, which lays its eggs inside the larval case and eventually destroys the inmate when it has reached the pupal stage. (Plate 45, 4)

The other type of caddis larva, the so-called *campodeiform* type (Gr. *kampe*, caterpillar), is nearly always found in running water, whereas examples of the eruciform larvae are found in both still and running waters. Campodeiform larvae rarely live in cases in the larval stage, but instead make a net attached to the underside of stones or on plants. In some species, as for example in *Hydropsyche*, many nets may be found together under a single stone. The open end of the net usually faces upstream, and small creatures are caught in the net and devoured. These larvae are more active than the case-building types, and usually have a flatter body, which tapers towards the rear end. The claspers are long, and some species have tufted tracheal gills on the abdomen, although others bear no gills, and presumably extract oxygen from the water by diffusion through their skin. At pupation time even this type of larva builds a case of stones, attaching it firmly underneath a large stone. The emergence of the adult caddis of swift streams is often extremely rapid, the fly coming out of the pupal skin as soon as it reaches the surface.

Some species of caddis-flies can produce scents or odours. These are of two kinds: scents from special structures near the mouth of the male, where dense tufts of hair unfold, and an odour of vanilla is given off; and a less pleasant smell given off by both sexes from a fluid excreted at the anal opening. It seems possible that the first kind of odour may be of value in mating, and it has been suggested that the second may serve to repel predators when the newly-emerged insect is particularly vulnerable. Species in which the first kind of smell has been noticed are *Sericostoma personatum* and *Halesus radiatus*. The second type of odour has been noticed in *Anabolia nervosa*, specimens of *Potamophylax* and *Limnephilus rhombicus*. Further observations would be valuable.

The eggs of caddis-flies living in still water are laid in gelatinous masses, which are either loosely attached to stones or plants above or in the water, sometimes in a closed ring, or dropped into the water as the females are flying over. Species which live in running water, however, take more care with their eggs, often descending into the water to lay them and fastening them securely to stones, to prevent them from being washed away by the current.

One hundred and ninety-three species of caddis-flies are recognized in Britain, and below are given the main families and details of the usual types of cases made by their larvae. (See also Plate 46)

PHRYGANIDAE   The larvae of this family are large, the cases of *Phryganea grandis*, the largest of our caddis-flies, measuring nearly 50 mm in length in a full-grown specimen. They are found in ponds and ditches, where there is an abundance of plant life, the cases being made largely from rectangular pieces of leaves wrapped spirally to form a cylinder. The eggs of *P. grandis* are laid in ropes of jelly entwined on submerged plant stems. (Plate 46, 1, 3G)

LIMNEPHILIDAE   The members of this family are the commonest caddis larvae and are found in ponds and still water generally. The many species of *Limnephilus* make their cases out of plant stems, stones, sand grains or snail-shells, and usually they are cylindrical in shape. (Plate 46, 3H) Those of *Potamophylax* (= *Stenophylax*) are made of sand grains (Plates 45, 9 and 46, 3A), while *Anabolia* has a characteristic case of sand grains to which are usually attached several long pieces of stick. (Plate 45, 1 and Plate 46, 3I)

SERICOSTOMATIDAE   Mainly running-water forms in which the cases are made of sand grains and curved, making them resemble a miniature hunting-horn. (Plate 45, 8 and Plate 46, 3F)

In *Goera* and *Silo* the central tube of the case has attached to it large pieces of stone. (Plates 45, 4, 6; 46, 3E) The adults of *Brachycentrus subnubilis* are the grannom or green-tail of the trout fisherman, and emerge in great numbers from some rivers such as the Kennet in Berkshire, in April.

MOLANNIDAE   The cases of *Molanna* which are found in lakes and larger ponds are easy to recognize, for they consist of a shield-shaped structure to which is attached a central tube, both parts being made of sand grains.

LEPTOCERIDAE   *Triaenodes* is probably the best-known genus of this family, with four species. The tapering case, about 25 mm long, is made of pieces of plant arranged spirally to form a slightly curved tube. These are often found in ponds with the larvae actively swimming with their hind pair of legs, which are well covered with hairs. At pupation time the cases are attached to plants. The eggs are laid in a disc of jelly on the underside of submerged leaves. The larval cases of the three species of *Athripsodes* are made of sand grains and are conical in shape, with a strong curve. (Plate 45, 2)

HYDROPSYCHIDAE   Most of the members of this family are found in swift streams during their larval and pupal stages. The larvae make no case, but build a silken net underneath stones, the opening to which faces upstream. At pupation time a flattish case of stones, loosely attached to a web of silk, is fastened to the underside of large stones.

POLYCENTROPIDAE *Polycentropus* is also an inhabitant of rivers or lakes, the nets often being found in large numbers clustered together.

Holocentropus occurs only in still water, the larvae being found under silken nets spun on the underside of large floating leaves of water-plants.

PSYCHOMYIDAE *Psychomyia* and its larger relative *Tinodes* occur at the margins of lakes and also in slow streams, making small tunnels lined with silk to which is attached plant debris or silt. The eggs are laid in masses attached to stones.

RHYACOPHILIDAE The larvae of *Rhyacophila* (Plate 46, 4) are found under stones in swift streams. They are active creatures, creeping about rapidly, and do not make a case until they are about to pupate, when they build a shelter of stones underneath a larger stone.

Agapetus is a very common inhabitant of streams, its stony case, flat on one side and rounded on the other, being found under large stones, many often being together. (Plate 45, 7)

HYDROPTILIDAE This family contains the smallest of the caddis flies, full-grown larvae being only about 5 mm in length. Until the larvae are a few months old they do not make cases and, having no gills, apparently breathe through their body surface. The cases made when the creatures are older are usually made of silk only and are not tubular, but flattened.

Aquatic Moths: Order LEPIDOPTERA

Only one or two species of this large order (Gr. *lepos*, a scale; *pteron*, a wing) comprising butterflies and moths, are aquatic and below the surface of a pond seems a strange place to find the caterpillars of moths, yet a few members of one family, Pyralidae, pass their larval stages under water.

They are commonly called the china mark moths from the fancied resemblance of the markings on the wings of some of them to the potters' marks inscribed on the bottom of good china. The presence, or otherwise, of the caterpillars of the brown china mark moths, *Nymphula nympheata*, in any pond or lake can readily be seen by examining the floating leaves of water-lilies and *Potamogeton* during summer. If present, the larvae will have cut oval pieces about 25 mm long out of the leaves, and on the underside of the leaves the missing pieces will usually be found attached by silk with a fat caterpillar between them and the leaf. The larvae are typical caterpillars with three pairs of true legs on the thorax and five pairs of prolegs on the abdomen. When first hatched they do not appear to suffer from getting wet and, as they possess no tracheal gills or other devices for breathing dissolved air, presumably they absorb what they need through their skin. After several moults, however, the body becomes covered with hairs, and if the case of an older larva is opened, the creature will be found quite dry inside, and is, no doubt, breathing the air with which the case is filled. When feeding, the caterpillar puts its head outside the case and nibbles the leaf to which it is attached. The aperture is so small and the pile of hairs so close that no water enters the case during this procedure. Pupation takes place in a silken cocoon, covered with pieces of leaf and attached to a plant stem just above the water-level. Specimens I have kept in aquaria have climbed out of the water and fastened the case to the glass sides of the tank at this time.

The adult china mark moths are on the wing among the waterside vegetation of ponds and lakes in June and July, the whole life-cycle apparently taking one year. The eggs are laid on the underside of floating leaves, the female bending her abdomen over the edge of a leaf to attach them in rows. The eggs hatch in about two weeks. (Plate 47, 1, 3)

A larger species, *Nymphula stratiotata*, although widely distributed, is commoner in southern England. The caterpillar feeds submerged on a variety of plants, including *Potamogeton*, *Elodea* and *Ceratophyllum*, and the larval period may occupy two seasons. The larva of this species is interesting in that its spiracles or air-holes are closed, and it possesses tuft-like tracheal gills with which it can extract oxygen from the water. Its case is not watertight like that of the commoner species, and the larva is constantly surrounded by water.

Pupation takes place under water in an air-filled cocoon spun by the larva and attached to submerged plants. The pupa has three pairs of spiracles, and of course breathes the air which is contained in the cocoon. The adult moths are on the wing in June, July and August.

The larvae of a smaller species, *Cataclysta lemnata* (Plate 47, 2), found on duckweed, have cases rather like those of caddis larvae, since they are made of several duckweed leaves fastened together. *Acentropus niveus*, a moth of another sub-family, also passes its larval stages under water, making a case like the others and breathing air through its skin. Pupation takes place under water. The adult moths are white and are on the wing from July. In this species there are two kinds of female moths, one fully winged and able to fly, whereas the other kind has only tiny wings and remains under water, where it is able to swim by means of its legs. It comes to the surface only to pair.

Two-winged Flies: Order DIPTERA

Diptera (Gr. *di-*, two; *pteron*, a wing) is a huge order of insects, which includes such well-known members as the common house-fly, the bluebottle, gnats, midges and horse-flies. None is aquatic in the adult stage, but a considerable number of them live in the water during the larval and pupal stages. Instead of the more usual four wings found in other insects they have only two, although in many cases the remnants of the lost pair are visible under a lens as tiny knobbed stalks, which are called halteres or, popularly, 'balancers', since if the insects are deprived of them they cannot control their flight.

No adult fly can eat solid food, and the mouth-parts are adapted for taking liquids. Many flies have a piercing organ to enable them to reach a liquid, in some cases the blood of an animal, in others plant liquids. The biters in such families as the gnats and mosquitoes are the females, for they alone have fully developed mouth-parts, and in some of them a blood meal is necessary before their eggs can develop.

The flies are divided into two main groups, depending on the form of pupa and the way the adult fly emerges from the pupal case. In the first group, Orthorrhapha, are included those flies in which the pupa is free, and the adult emerges through a slit in the thorax of the pupal skin. In the other group, Cyclorrhapha, the larva is of the type usually called a maggot, and when the time comes for it to pupate the larval skin, instead of being cast as is more usual in insects, remains as a hard protective case, or *puparium*, in which the maggot changes into a pupa. The adult fly emerges from the puparium by pushing off one end with the help of a tiny bladder, which can be protruded through a slit in the head and is inflated by the pressure of blood.

It is with the first group of flies that we shall be more concerned, since a great many of them live in water in their larval and pupal stages, and the eggs of some are laid on or in the water. Only two families of the second group, Syrphidae, hover-flies, and Anthomyiidae, close relatives of the common house-fly, have aquatic members.

The first group is further split into two sub-orders, Nematocera (Gr. = thread-horns, from the long, many-jointed antennae of the adults) and Brachycera (= short-horns). In the former the larvae have well-developed heads with biting mandibles and antennae, but in Brachycera, the larva has only an apology for a head, which can usually be withdrawn into the first segment of the body.

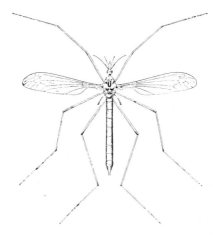

Fig. 75   Crane-fly, 25 mm

o

Sub-order NEMATOCERA ('Thread-horns')

TIPULIDAE   Adult crane-flies or 'daddy-long-legs' (Fig. 75) are familiar insects and the larvae of some of them are the notorious leatherjackets, which do so much damage to growing crops. The females lay several hundred eggs from June onwards, inserting their long ovipositors almost vertically into the ground. The larvae hatch in about a fortnight, but they do not pupate until the following spring or summer. Several species pass their larval stage in the damp earth near the water, and others actually in the water. They are worm-like creatures, yellowish white in colour, and when full-grown may be 40 mm or more in length, with a number of characteristic tube-like outgrowths on the last segment of the body.

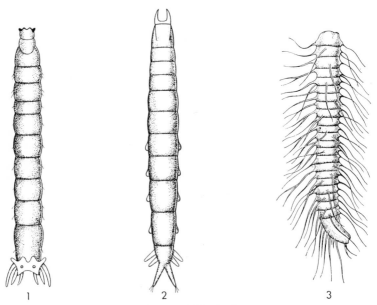

Fig. 76   Fly larvae: 1 *Tipula*, 40 mm.
2 *Pedicia*, 40 mm. 3 *Phalacrocera*, 30 mm

The larvae of the genus *Tipula* can be recognized by a plate-like structure at the rear end bearing six or eight lobes which surround spiracles. In shallow water the larvae hang from the surface, using the plate as a float and exposing the spiracles to the air, but when submerged the insects fold the lobes over to enclose a bubble which may assist respiration. The larva of *Dicranota*, which is smaller and more active, has five pairs of prolegs on the abdomen and is usually found on the bottom of streams where it searches out and eats *Tubifex* worms and similar fare. The large larva of *Pedicia*, up to 40 mm long, has four pairs of prolegs on the abdomen, and constantly inflates and deflates the penultimate segment of the body, presumably for respiratory purposes.

Pupation takes place in more or less vertical burrows in the muddy bank and the pupae are smaller than the larvae. In *Tipula* pupae there are long horns, believed to be for respiration, on the first segment of the thorax. In *Dicranota* the horns are shorter.

The larva of *Phalacrocera replicata*, which is about 30 mm long, is quite unmistakable. It has long processes arising from its body, which make it very difficult to see among the sphagnum moss in which it lives, especially as its colouring of green or brownish resembles that of the plant.

PTYCHOPTERIDAE   The adults of *Ptychoptera* are also called crane-flies, but the larvae, which may be 70 mm long, are quite different from those of the tipulids, having a long, slender, tail-like structure at the end of their elongated body. The head is distinct, and where the tail joins the body are two small

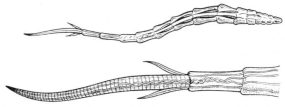

Fig. 77 Larva of
Ptychoptera, 70 mm.
Below: detail of
rear end of body

appendages. The larvae (Plate 50, 6) spend most of their time buried in the mud at the bottom of shallow ponds with their long tail, which is really a breathing-tube, reaching to the surface of the water. The tail, which can be lengthened or shortened to suit the depth of water, has two tracheal tubes running through it which communicate with the air through two openings at the extremity.

The pupae live on the mud, but instead of a respiratory tail have, on the head, a long tube which reaches the surface, and air is taken in through the thin walled end. There is also a second much shorter tube on the head.

PSYCHODIDAE   Some of the tiny moth-flies are aquatic, and two species at least have become known as 'sewage-flies', from the successful colonization by their larvae of the bacteria beds of sewage farms, where they feed on algae. They occur in such numbers that they are an important check on the growth of the algae, which would otherwise tend to choke the filter beds and prevent them from carrying out their function of purifying the sewage effluent. Unfortunately, however, the immense swarms of adult moth-flies are not regarded with any enthusiasm by people living in the neighbourhood of sewage farms, and some check on the numbers of the insects has to be enforced.

DIXIDAE   The larvae of the *Dixa* midges are very common at the surface of weedy ponds and streams, and they are readily distinguished by the way they keep bending their body into a U-shape. The full length of the body is only about 5 mm. Around the mouth are tufts of bristles, and the vibration of these causes currents of water which bring microscopical food particles to the mouth. The pupa rests at the surface, and its body is similarly bent into a U-shape.

CULICIDAE   The adults of this family are the best known of all aquatic Diptera, for they include the gnats and mosquitoes which are such a plague on humanity and which, through being hosts of a number of man's parasites, rendered large parts of the earth's surface almost uninhabitable until modern methods were found of controlling them. It has already been mentioned that it is the female only which has piercing organs in the form of mandibles, modified into needle-like stylets. Fortunately, the large swarms of gnats seen on summer evenings consist solely of males and present no danger to humans. The females wait on nearby vegetation until their time comes for mating, when they fly into the swarm and pair quickly with the first successful male. The pair drop to the ground, the whole affair lasting only a few seconds. The male is an innocuous creature which can be distinguished at sight by his beautiful feathery antennae, in contrast to those of the female, which bear only small hairs. The two best-known genera of the mosquitoes are *Culex* and *Anopheles*. It is important to differentiate between them, because some species of *Anopheles* are capable of transmitting malaria, while *Culex* is not, although some species often attack

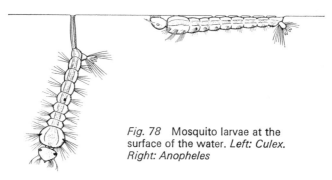

Fig. 78   Mosquito larvae at the surface of the water. *Left: Culex. Right: Anopheles*

humans, and even cause very painful bites. The adults are easily distinguished by the attitude they adopt when resting on a surface. Whereas *Culex* holds the body roughly parallel to the surface, *Anopheles* is inclined at an angle with the head almost touching it.

The larvae, too, take up different attitudes in the water, those of *Culex* hanging downwards at a slight angle to the vertical, the tail just touching the surface film, while anophelines rest with the length of their bodies along the surface. *Culex* larvae take in air through a long breathing-tube, whereas those of *Anopheles* do so directly through spiracles on the eighth abdominal segment.

Further differences are seen in the method of egg-laying. The eggs of *Culex* are laid in masses or 'rafts' floating on the surface; those of anophelines are laid singly, with a small float on each side to make them buoyant in a horizontal position.

Twenty-six species of culicine mosquitoes and four species of anophelines occur in Britain, all passing their larval and pupal stages in a variety of still-water habitats, such as ponds, water-butts, brackish pools and even in the water which collects in holes in trees.

Culex pipiens is perhaps the commonest of all, breeding in almost any small open area of water, including water-butts and stone garden troughs. The cigar-shaped eggs, numbering about 300, are cemented together into the form of small boats or rafts about 5 mm long, which float with the eggs broad end down-wards, on the surface of the water. They are most easily found at the sides of water-butts, and once one becomes accustomed to seeing them, it is surprising how many will be found throughout the summer. The rafts are unsinkable, and no matter how long they are forced under water, they will bob up to the surface quite dry. At the bottom end of each egg is a little trap-door, through which, in due course, the legless, wriggling larva escapes into the water. It spends most of its life suspended from the surface film by the tip of the long breathing-tube situated on the eighth segment of the abdomen. Five tiny flaps at the extremity of the breathing-tube open out into a kind of funnel when the creature is at the surface, and allow air to reach the spiracle which they surround. The last seg-ment of the body has four gill-like structures, which are believed to be for the purpose of regulating the salt content of the body.

By means of vibrating brushes on their heads, gnat larvae create currents in the water which bring to the mouth the minute particles on which they feed. In about three weeks the larvae change to pupae, which are shaped like a comma with a large dot. In this stage they do not feed, but pass their time resting at the surface, taking in air through two tiny 'ear trumpets' which arise from the thorax just behind the head. Eventually the pupal skin splits where it touches the surface film, and the perfect insect escapes into the air. (Plate **49**)

C. pipiens rarely bites human beings, confining its attention mainly to birds. Bites which have been attributed to it in the past are now believed to have been inflicted by a very similar species, *Culex molestus*, whose presence in this country was unsuspected until 1934. A much larger culicine, *Theobaldia annulata*, which is easily recognized by its speckled legs and dark spots on the wings, and which sometimes breeds in garden pools, but which prefers waters rich in organic matter, is a very vicious biter, and most cases of really bad gnat-bite are probably due to this insect. The fourteen species of *Aedes*, whose larvae resemble those of *Culex*, are also troublesome biters of human beings and are very common.

The mosquito which is capable of transmitting malaria in northern Europe, including the British Isles, is *Anopheles maculipennis*, but it has been found that there are three different varieties of this species, which are almost indistinguish-able in the adult stage, but have certain differences of behaviour. Two of these varieties, *messeae* and *typicus*, however, confine their attentions almost entirely to domestic animals, but *A. maculipennis* var. *atroparvus*, which breeds in brackish water, attacks man. In a coastal area with a source of malarial parasites, such as people who have lived in malarial areas and had the disease, it could infect human beings. Malaria, or ague as it was called, was a common illness in Britain not very long ago, particularly in areas such as the Fens of eastern England, where conditions were favourable for anopheline mosquitoes. It is on record that in some summers the harvest was lost in various parts of the country, even as far north as Inverness, because of the prevalence of ague among the labourers.

The great popularity of garden pools in recent years, even in small gardens, has prompted questions as to their danger as potential breeding-places for mosquitoes. It is clear, however, that the possibility of malaria-carrying species breeding in them is remote, and the almost inevitable presence of goldfish or other fish in such pools is a sufficient guarantee that such mosquitoes as do choose to breed there will rarely survive to the adult stage.

In passing, it is of interest to note that one culicine larva has adopted a novel method of breathing. Instead of taking in air at the surface, *Taeniorhynchus richiardii* inserts its breathing-tube, which is rather sharply pointed, into the roots of water-plants and taps the air space. The pupa also obtains its air supply from the same source, pushing its ear trumpets into the root tissue of plants.

In addition to the mosquitoes, the family *Culicidae* also includes the insects sometimes called the phantom midges, although they are not really true midges. The larvae are the well-known and well-named phantom or glass larvae of the four species of *Chaoborus* (formerly *Corethra*), which can be found not only in small ponds but also among the plankton of large lakes. About 10 mm long, they are quite transparent and very difficult to see in a collecting-tube. They remain motionless in the water, in a horizontal position, for long periods, and then with a flick of the body they vanish only to reappear some distance away. The two pairs of dark sausage-shaped bladders, one at each end of the body, contain a gas slightly different in composition from air and enable the creature to maintain its equilibrium in the water. The bladders are distended when the larva rises in the water and contract when it is going down. At the tail end of the body there is a very beautiful comb of bristles which may act as a rudder and stabilizer. There are no spiracles, and respiration must be carried out by gaseous exchange with the surrounding water through the skin.

The adult midges are not blood-suckers. Females lay their eggs in flat discs of gelatinous material with the eggs, up to about one hundred, arranged in a spiral. (Plate 48, 3)

Phantom larvae are carnivorous, using their long modified antennae as grasping organs with which to capture the small creatures on which they feed. They remain as larvae over the winter months, swimming about in the coldest weather, and have even been thawed out of blocks of ice without showing any ill effects. The pupae are similar to those of the rest of the family, with large heads and breathing-trumpet. The adult flies emerge in spring.

SIMULIDAE   The black-flies (sometimes incorrectly called sand-flies) are not sufficiently troublesome in Britain to have attracted much notice, but in other parts of the world, under such names as buffalo gnats, they are only too well known as pests of cattle, the pain and discomfort caused by their mass biting attacks driving cattle and horses to a frenzy.

The females of British species do bite a variety of birds and mammals, including human beings, and in Scotland, *Simulium tuberosum* has become sufficiently notorious to attract the common name birch-fly.

The family comprises three genera containing thirty-five species. The larvae are up to 10 mm long. The times of appearance of the four stages vary with the species. Some over-winter as eggs or larvae, and pupate in early summer. Others are on the wing as adults from April to June.

The eggs are laid in masses of jelly on stones and plants at the edges of streams, and the larvae enter the water and attach themselves to similar supports, usually in the most rapid part of the stream. Sometimes the plants are black through the vast number of larvae covering them. The method adopted by the creatures of attaching themselves to their support is both interesting and effective. The surface is first covered with a secretion, and the larvae then hold on to that by means of a ring of hooks at the tail end of the body. If a larva becomes dislodged from its support, it quickly throws out a line of mucus, in the manner of a spider's thread, and is thus prevented from being washed downstream. Food particles are sieved out of the current by two prominent combs of bristles on the head, and are then conveyed to the mouth. A form of tracheal gill on the thorax enables dissolved air to be utilized for breathing purposes.

When the larva is full-grown, it spins an open brown cocoon, like an ice-cream cornet, attaching it to the support on which it has lived, and then pupates. (Plate 50, 5) Protruding from the cocoon can be seen the two tufts of thoracic breathing-filaments with which the pupa obtains its supply of oxygen from the water.

The manner in which the adult *Simulium* fly emerges into the air is surely the most wonderful method of all those used by aquatic insects. The pupa absorbs from the water more air than it needs for breathing, and this air accumulates in the space between the outer skin of the pupa and the skin of the developing adult inside. When the time comes for emergence, the pupal skin splits and the adult fly floats to the surface enveloped in a silver bubble of air. The wings are ready to function as soon as the fly emerges, and it takes to flight immediately it reaches the surface.

CHIRONOMIDAE   The true midges are a large family of twenty-six genera, comprising some four hundred species, most of which cannot be differentiated in the larval stage. Their total population in many freshwater habitats is immense. How fortunate it is that they are not biting insects!

In summer evenings the males may be seen in vast clouds over water and elsewhere, looking like a plume of smoke from a distance, and sometimes causing alarm when they congregate over the steeple of a church or other similar place. There is scarcely a freshwater habitat that does not contain the larvae of one or other species, for of all fly larvae they are among the most completely adapted for an aquatic life. By reason of their abundance and ubiquity, they are one of the most important sources of food for other aquatic creatures.

Some of the larvae of *Chironomus* are the familiar 'bloodworms' which occur in a wide variety of environments and are about 15 mm long when full-grown. Near at home, rain-water butts and garden troughs are sometimes teeming with them, and the small shelters or burrows of silt, attached to a supporting mesh of a silky secretion, will be found adhering to the sides. At times, the larvae leave the shelter and swim rapidly through the water by means of vigorous contortions of the body. (Plate 48, 1)

The red colouring of the body of some species is due to the presence in their blood of haemoglobin, the pigment which is found in human blood and that of all the higher animals. This substance has, of course, the property of combining with oxygen very readily to form oxyhaemoglobin, which rapidly gives up its oxygen to any tissues of the body which need it. It has been shown by experiment that the haemoglobin in the blood of chironomid larvae is of value to the creatures in conditions where oxygen is scarce, but that in waters adequately oxygenated it is apparently not used. Their possession of haemoglobin, therefore, enables bloodworms to live in conditions under which other creatures would be unable to obtain sufficient oxygen, and this accounts for their presence in the muds of deep lakes or of stagnant pools rich in decaying organic matter.

Species which live near the surface and do not need such a provision are pale in colour and sometimes greenish. Respiration in all species takes place through the general body surface which is richly supplied with fine trachea. The so-called blood-gills on the last but one segment of the abdomen take no part in respiration, as was once thought. They are for the absorption of salts from the water, to maintain the ionic composition of the blood.

The food of chironomids consists mainly of particles of organic matter obtained from the detritus of the bottom. Some species spin funnel-shaped nets at the entrance to their burrows in the silt and then, by rhythmic movements of the body, create currents in the water which draw material to the nets. When these are full the larvae eat the net and its contents. In lakes, most larvae take two years to complete their development.

The pupae resemble those of the mosquitoes in general form, but instead of breathing-trumpets on the head, they have tufts of filaments which are capable of absorbing dissolved oxygen from the water. The pupa swims to the surface to enable the adult midge to emerge.

The eggs are laid in ropes near the surface of the water, usually attached to a plant or other object. In water-troughs and similar places the ropes, about 25 mm long, will be found fastened to the sides, usually a few centimetres below the surface. (Plate 48, 2)

Tanypus larvae are common in ponds and resemble those of *Chironomus* in general form, but are smaller and lack the red colouring; they are carnivorous. The eggs are laid in gelatinous blobs on submerged stones or plants.

CERATOPOGONIDAE The midges described above are not troublesome insects to man, for none of them is able to pierce his skin with its mouth-parts. The females of the biting or blood-sucking midges belonging to the present family, however, can be and are, in fact, considerable nuisances. Although only about $1\frac{1}{2}$ mm long, their numbers and persistence make many parts of the country most unpleasant places to be on a summer's evening. *Culicoides*, for instance, is particularly active in the Western Highlands of Scotland. Midges are so small that they can penetrate between the threads of clothing and stockings with their piercing mouth-parts, and some people seem particularly sensitive to their bites.

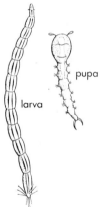

Fig .79 Ceratopogonid midge: larva, 4 mm, and pupa

In some species the larval and pupal stages are passed in or near running water, but most are found in association with stagnant water, the larvae feeding on algae. The larvae are about 4 mm long, elongated and worm-like, and swim in an undulating way. Their development is very rapid, and the adult midges are on the wing three weeks after the eggs have been laid.

Sub-order BRACHYCERA ('Short-horns')

STRATIOMYIDAE The soldier-flies are the first of the stouter, short-horned flies to be considered; their name dates from times when military uniforms were more resplendent than they are today. The adults of the genus *Stratiomys* are somewhat sluggish insects, about 12 mm long, seen resting on waterside vegetation in summer. An example is *Stratiomys chamaeleon*, often called the chameleon-fly, which although not the commonest species in Britain, is well-known through the memorable accounts of it written by Swammerdam and

Reamur (p. 3), which have since been repeated in many well-loved natural history books.

Their broad flat bodies have a ground colour of velvety black, with broad yellow markings, and they have a bee-like appearance which is accentuated by their habit of sucking nectar from flowers in the vicinity of ponds and ditches. The females lay their eggs in the leaves of marginal plants, and the strange-looking legless larvae make their way into the water on hatching. They are tough-skinned, long-bodied creatures, tapering towards the tail end, brownish or greenish in colour and about 35 mm in length when full-grown.

They spend most of their time hanging from the surface film by means of a tuft of about thirty branched filaments at the rear end of the body. These open out when the larvae are at the surface and bring the two spiracles at the end of the body into communication with the air. When the insects go down into the water, the filaments close to surround an air bubble which provides them with an air supply while submerged. (Plate 50, 4)

The head is small and capable of being telescoped into the body, and the mouth-parts are kept in constant motion to bring to the jaws the minute creatures on which the larvae feed. The pupa floats horizontally at the surface and is remarkable in that it is formed within the apparently unchanged larval skin. It is, in fact, difficult to say whether a larva or pupa is being studied, but a close examination at the pupal stage will reveal the pupa occupying a small space near the head end of the case, the rest of the space being filled with air to provide for the needs of the pupa. The outer skin is very hard at this stage. The adult fly emerges through an opening near the head end of the case.

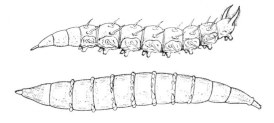

Fig. 80   Above: larva of Atherix,
15 mm.
Below: larva of a tabanid,
20 mm

RHAGIONIDAE   The remarkable larvae of the three species of *Atherix* are occasionally found on the stony bottoms of rivers, but they are not common. They are about 15 mm in length and have characteristic fleshy pseudopods bearing spines on each abdominal segment, well-developed prolegs and two long appendages at the rear end with long hairs. The females cluster on water-side vegetation when egg-laying and then die, so that a mass of dead flies and eggs is found.

TABANIDAE   Horse-flies and clegs are well enough known to most people on account of their painful 'bite', and they are also serious pests of cattle, for their attacks disturb grazing herds, making them restless and unable to settle down to feed properly. It was formerly believed that the 'gadding' of cattle—that is, the mad rushing about with tail aloft and neck outstretched—was due to these insects, and some were, as a consequence, called 'gad-flies', but it is now known that this behaviour is caused by quite a different pest, the warble-fly, *Hypoderma*.

The horse-flies are large insects, resembling a big house-fly in general form. The female flies alone have the fully-developed piercing mouth-parts which are capable of penetrating the skin and sucking the blood, the males contenting themselves with sipping water or nectar.

The larvae of some of the twenty-eight British species of this family live in or near water, or on swampy ground, frequently among damp moss. They are legless, grub-like creatures about 20 mm long when full-grown. Their bodies are soft and pale-coloured, tapering to each end, while on the seven middle segments are circlets of fleshy protuberances which serve as legs. At the tail end of some of the aquatic species is a breathing-tube which is protruded out of the water when the creature comes to the surface to breathe. They feed on worms or insect larvae. Pupation takes place in damp earth near the water, and the females in due course lay their egg-masses on the stems of waterside plants.

The true horse-flies belong to the genus *Tabanus* and the clegs to the genus *Haematopota*. Members of the genus *Chrysops* are often called deer-flies, presumably because they frequent the deer forests of Scotland.

The various members of this family have different methods of attack. Clegs usually bite on the wrist and give no warning of their approach; horse-flies more frequently alight on the legs and ankles, making a loud buzzing noise when approaching their victims, and the deer-flies, after a preliminary high-pitched buzz, settle on the back of the neck.

Sub-order CYCLORRHAPHA

It was mentioned earlier in the chapter that the only families of Diptera with aquatic representatives belonging to the second great group of flies, in which the larva is in the form of the maggot and the pupal stage passed in a hard protective puparium, were Syrphidae and Anthomyiidae and, for the sake of completeness, brief mention must now be made of them.

SYRPHIDAE   The hover-flies and drone-flies are the large bee-like insects often seen hovering over flowers, and one has to look closely to make sure that the creatures are not really four-winged bees or wasps. (Plate 50, 3) The larvae of some of them live in water, the best-known being *Eristalis*, the familiar rat-tailed maggot, found in the black mud in the shallow parts of small ponds rich in the decomposing matter on which it feeds. To find this creature it is best to scoop up a quantity of the evil-smelling mud, spread it thinly on the bottom of a large shallow dish and just cover the mud with water. When the sediment settles, any rat-tailed maggots will easily be seen. The actual body of the creature is only about 15 mm long and of a greyish colour. The end of the body, however, is extended into a long breathing-tube made up of three segments, which can be telescoped into each other. This normally reaches the surface of the water and can be made longer or shorter as the creature pleases to suit the particular depth at which it finds itself. The tip of the tube is surrounded by feathery hairs, in the middle of which are the spiracles. (Plate 50, 1)

The larva might be confused with that of the crane-fly, *Ptychoptera* (p. 201), but the latter can readily be distinguished by its small but distinct head, that of *Eristalis* being almost non-existent.

The pupa of *Eristalis* remains within the last larval skin and retains the same outward appearance, except that the skin has contracted a little, and now bears two small breathing-horns with which the pupa, in those species which remain in the water at this stage, takes in air while floating passively at the surface of the water. Some, however, leave the water and pupate in the damp ground near by. (Plate 50, 2)

The eggs of *Eristalis* are laid in small clusters on the surface of the water.

MUSCIDAE This large family includes the house flies and a large number of garden and agricultural pests. A number of maggot-like larvae are found in fresh waters, but little is known about their identification. The larva of *Limnophora*, however, is distinctive with a long, tapering anterior end and an oblique hind end. It measures about 10 mm long and is usually found in running water.

Ichneumon Flies and Chalcid Wasps: Order HYMENOPTERA

The order Hymenoptera (Gr. *hymen*, a membrane; *pteron*, a wing) is a large one, including the sawflies, ants, bees, wasps and ichneumon flies, but only a very few of its members are aquatic or semi-aquatic. These comprise one or two species of tiny insects which parasitize various aquatic creatures. One of these, *Agriotypus*, has already been mentioned (p. 195) as a parasite of caddis larvae. The adult insect is a small ichneumon fly, the females of which crawl down into the water and lay their eggs inside the larval cases made by the Phryganidae family. They carry an air film which prevents the wings being used under water, and the antennae are pressed back against the dorsal surface so that the tarsi of the legs have to be used to locate suitable caddis cases in which to lay the eggs. The ichneumon larvae feed on the tissues of the caddis larvae, but do not kill them. Evidence that *Agriotypus* has been at work is revealed by the long filament which is attached to one end of any cases they have occupied. (Plate 45, 4)

Related insects belonging to the family Braconidae similarly parasitize larvae of the two-winged flies.

Two genera of chalcid wasps are also aquatic. *Caraphractus cinctus* (= *Polynema natans*), an insect only about 1 mm long, lays its eggs singly inside those of water-beetles, such as *Dytiscus*, *Ilybius*, *Agabus*, *Colymbetes* and *Hydroporus*. The tiny larva eats the egg contents, and then pupates in the empty eggshell, emerging as an adult about a fortnight afterwards. The adults have a waterproof surface, but carry no obvious air film. Nevertheless, they crawl about on submerged water-plants, and even swim in a jerky fashion, using their delicate wings to propel them through the water. Specimens which have been under observation have been kept submerged as long as twelve hours without showing any need to obtain a fresh supply of air at the surface. (Plate **48, 4**)

A related insect, *Prestwichia aquatica*, lays its eggs in those of beetles, including *Dytiscus* and *Pelobius*, and bugs such as the water-boatmen, *Notonecta*, and the water scorpion, *Ranatra*. It has a similar life-history to *Caraphractus*, but in swimming it uses its legs to propel it through the water. The males of this insect are wingless and the females have only small wings, which are useless either for flying or swimming.

Diapria parasitizes fly larvae, including those of the rat-tailed maggot, *Eristalis*.

16: THE ARACHNIDS

The last group of arthropods to be considered, the class Arachnida (Gr. *arachne*, a spider), is a large one containing, among other creatures, the spiders, scorpions, mites and harvestmen. Comparatively few, however, are aquatic, and these all belong to two only of these orders—Araneae, the spiders and Acari, the mites.

There is a common misconception that these creatures are insects, but even a cursory examination would reveal several very obvious differences. Thus adult arachnids have four pairs of legs instead of the three pairs typical of insects; they have no wings; their head bears no antennae; and the front part of the body is not divided into head and thorax. There are other more subtle differences. Instead of the three pairs of mouth appendages generally found in insects, arachnids have only two, and their eyes, usually eight in number, are simple with one lens in each; they never have the elaborate compound eyes bearing many facets which are so characteristic of most insects.

The Water Spider

Although many kinds of spiders frequent the waterside, only one, the water spider, *Argyroneta aquatica*, lives actually in the water. This interesting creature is quite common in ponds and still water generally in many parts of the country. The females are about 10 mm in length, the males usually slightly larger—an unusual feature in spiders, for the female is nearly always the larger. No doubt owing to this male superiority, water spiders have apparently achieved a state of marital bliss, and the female does not eat her mate after pairing as so frequently happens in the spider world.

In appearance the water spider is similar to land spiders usually encountered in home or garden. The division of the body into the *prosoma* and oval-shaped *opisthosoma* or abdomen, and their separation by a wasp-like waist, is typical of the order *Araneida* to which they all belong. The four pairs of jointed legs are all borne on the fore-part of the body, and what at first sight might be taken for antennae are seen, on closer examination, to be a pair of long-jointed jaw appendages, which are called *pedipalps* and serve as touch organs. The front pair of fang-like jaws or *chelicerae* are much smaller, and are not readily seen from above, as they are hidden by the head. They consist of only two joints, a sharp-pointed end joint, which can be closed on to the grooved basal joint, rather like the blade of a penknife closing into its handle. The basal joint is furnished with a poison gland by which prey can be paralysed.

The abdomen bears no appendages except the finger-like spinnerets at its tip. From the many minute pores of these is exuded, in a liquid form, the web-making silk.

Spiders have tracheal or breathing-tubes within their body similar to those of insects, except that they are not branched. Spiracles on the underside of the abdomen enable air to be taken into the body and carbon dioxide to be given out. Some spiders, however, including *Argyroneta*, have also in their abdomen a pair of remarkable breathing-organs popularly called 'lung books', since they consist of a series of flat plates rather like the leaves of a book. These communicate with the outside of the body by means of another pair of slit-like spiracles on the abdomen. In the spaces between the plates there is some of the colourless blood, and air taken in at the spiracles is thus absorbed by the blood and circulated to other parts of the body.

Although the water spider has adopted an almost exclusively aquatic existence, it still has to breathe atmospheric air, and its methods of achieving this make it of great interest. These methods can easily be studied in a specimen kept in an aquarium.

When a water spider is first placed in water the whole of its abdomen takes on a silvery appearance by reason of the large bubble of air trapped in the fine velvety pile of hairs. This supply is adequate to supply the spider's needs for some time when submerged, and it can be renewed by occasional visits to the surface.

For permanent residence below the surface, however, a more elaborate means is required, and the water spider has adopted quite the most effective method of any aquatic creature—an air-filled shelter in which it can live, if necessary, for months.

The first stage in the construction of this is the spinning of a platform of silk anchored to water-plants. Then, by swimming or climbing up water-plants, the spider makes repeated journeys to the surface, each time returning with a bubble of air trapped between the legs and abdomen. By a stroking movement of the legs the bubbles are released underneath the web, pushing it up, and in due course the amount of air is such that a bell-shaped space is formed. In this the spider can live and breathe atmospheric air without the necessity of coming to the surface. It does not always enclose itself completely, but may spend long periods with just the abdomen inside.

Once the bell is made, the air supply does not need frequent replenishment from the surface, for as the oxygen becomes used up by the breathing of the spider, more oxygen will diffuse in from the surrounding water; similarly, the carbon dioxide breathed out will diffuse outwards. Thus the bell is self-acting and much more efficient than seems apparent on first consideration. (Plate 51, 1 and 2)

From the shelter of the bell, occasional sorties are made, usually along the lines spun between adjacent objects, in search of the live or dead creatures which form the spider's food. These are almost invariably taken back to the bell and consumed there.

In summer the female lays fifty to a hundred eggs and encloses them in a white egg-sac at the top of her bell, afterwards partitioning this portion off. Spiders do not pass through a larval stage, and the young spiders, when they hatch, after two to three weeks, are similar in appearance to the adults, but with incomplete mouth-parts.

When winter approaches, water spiders make their bell near the bottom of the pond, and remain in it in a torpid condition until spring. The young often use the empty shell of a snail as a shelter, spinning a web in it and filling it with an air bubble, as a result of which it may float to the surface.

If water spiders are kept in aquaria, it is as well to keep only one, or at most a true pair, in each receptacle, as they show cannibalistic tendencies. Dead flies placed on the surface, and *Daphnia* or similar live food in the water, will supply their needs for food. As the spiders are not altogether dependent on air dissolved in the water for breathing, quite small vessels, such as clear glass tumblers, are adequate and have the advantage of enabling a closer study to be made of the creatures than if they were kept in larger aquaria. A really close-fitting cover is, however, essential, as water spiders are adept at escaping through the smallest crevice.

Some of the wolf spiders belonging to the family Lycosidae live on marshes and are recognized by the two rows of small white spots down the abdomen and the white bands along the sides of the prosoma. They include *Pirata piraticus*, *P. hygrophilus* and *P. piscatorius*. The last spins a vertical silken tube among sphagnum moss. From the upper end the spider darts out to catch passing insects. The lower end of the tube is below the surface, and through it the spider passes when alarmed, a silvery bubble of air encasing the body and legs.

Dolomedes fimbriatus, belonging to the family Pisauridae, is our largest spider, measuring up to 22 mm in length; it has two yellow or cream stripes along the dark brown body. The common name, raft spider, is based on the mistaken belief that the spider floats downstream on a raft of leaves tied with silk. In fact, it lives at the edge of still waters in dykes and swamps, resting on the leaves of water-plants, with the front legs touching the surface of the water in a good position to detect ripples caused by insects which have alighted. The prey is caught after a quick dart across the surface and the spider then returns to its original resting-place along a thread laid down for the purpose. When alarmed, the spider runs down a plant stem into the water, the body and legs covered in a silvery bubble of air.

The Water Mites

The water mites, which form the small sub-order Hydracarina (or Hydrachnellae) (L. *acarus*, a kind of mite), occur in most stretches of fresh water, some swimming actively in the water or clambering about water-plants, whereas others are more sluggish and crawl about on the bottom.

They are similar in appearance to very small spiders except that their bodies are not divided into two distinct parts, but are in one piece, oval or round according to the species. Most of them are brightly coloured, reds predominating, but green, brown or even bluish species occur. They vary in size from less than 2 mm to a maximum here of about 8 mm. (Plate 52)

To study the structure of a mite properly it is necessary to turn it on its back, and examine it under a low power of the microscope. It will then be seen that the four pairs of hairy legs are attached to broad plates on the body called *epimera*, which may be separate or fused together. The particular form and arrangement of the epimera are important identification features in some species. The legs terminate in claws, and in the kinds which swim actively about there are, in addition to the usual fine hairs, long swimming-hairs near the joints.

Between the first pair of epimeral plates is a large complex structure called the false head or *capitulum*, containing the pointed beak with which a digestive fluid is pumped into the prey and the liquified food sucked into the body. Here also is the single pair of palps. The two pairs of eyes, and sometimes a single middle eye, will be noticed at the forward part of the body. Although fairly large, the eyes, judged by the apparent indifference of the creatures to bright illumination under the microscope, do not seem to be very sensitive to light.

Respiration is carried out through spiracles and a tracheal system, but the spiracles are covered with a thin membrane and gaseous exchange takes place through this from the surrounding water, making visits to the surface unnecessary.

Most species of water mites are very active little creatures, swimming rapidly and steadily on a straight course or clambering about among water-plants. Even in winter they will often be found bustling about in the water. They are carnivorous, and attack and devour small creatures they encounter, but quite often they seem to bump into them by accident rather than to seek them out. Their prey includes water fleas, chironomid larvae, worms and sometimes other mites. The adults of one genus, *Unionicola*, are parasitic on mussels.

The females of many species deposit their eggs on submerged stones or plants, usually enveloped in a gelatinous covering which in time hardens and serves as a protection, not only against predacious creatures, but also probably against desiccation should the pond dry up. Other species insert their eggs into punctures made in the stems of water-plants, and even in one case inside the shell of a freshwater mussel.

The larvae are unlike the adults in having only three pairs of legs. The false head is also proportionately larger. The larval stage of most mites is passed as a parasite on some other aquatic creature, usually an insect. Frequently, when one is handling a water scorpion, *Nepa*, or a *Dytiscus* beetle, a number of these tiny

P

red specks will be found attached to various parts of the body and legs, with their suctorial beaks embedded in the tissues of their host, obtaining all their food from the body fluids of their victims. (Plate 52, 4)

After hatching, the mite larvae have a brief free-swimming period and then they become attached to their hosts, remaining thus for a variable period before pupating, still attached to the host. Soon an active nymphal stage (the nymphopan) emerges from the pupal skin, escapes into the water and seeks its own food. After a time the nymph attaches itself to a solid support by its mouth-parts and enters a second pupal stage (the teleiophan). In seven to ten days the sexually mature mite emerges.

Some species of *Limnesia* and *Piona*, living in streams, do not have a parasitic stage but remain in the gelatinous egg mass until they change into nymphs.

There are over 200 species of British water mites belonging to about twenty families, and a few of the genera most frequently encountered are illustrated on Plate 52. These and others are described below:

Limnochares aquatica is unusual in having a rather long, rectangular-shaped body and being sluggish in its movements. It is usually found crawling about on the mud surface. (Plate 52, 2) Various species of *Hydrachna* are common in still water, and their larvae are the ones most frequently found attached to water-bugs, sometimes as many as several dozen of the little creatures being on one host. The globular body of the adult is dark red and the legs are short, with the exception of the front pair, which are well supplied with swimming hairs. The females deposit their eggs in holes pierced in water-plants. (Plate 52, 2, 4)

Another very common species is *Hydrodroma despiciens*, about 3 mm in length and bright red in colour. (Plate 52, 1)

The species of *Hygrobates* are somewhat oval in shape and usually greenish or bluish in colour. Some are very common in ponds covered with duckweed, *Lemna*, on the fronds of which the eggs are deposited. (Plate 52, 5, 10)

The species of *Atractides* are recognizable by the reddish T-shaped marking on the back. (Plate 52, 6) Species of *Limnesia* are common in slow rivers and are found especially in spring. (Plate 52, 7). *Arrhenurus* species are green or brown in colour, and the differences of shape between the male and female are shown in Plate 52, 8, 9.

Unionicola bonzi is found inside freshwater mussels, and lays its eggs there; the larva of another species, *U. crassipes*, sometimes occurs as a parasite on the freshwater sponge, *Spongilla*. *Mediopsis orbicularis* has a flattened, circular body and its eyes are borne on the extreme edge of the forepart of the body.

Water Bears: Class *TARDIGRADA*
(L. *tardus*, sluggish; *gradus*, a walk or step)

The water bears were, formerly, included with the arachnids, although, beyond the possession of four pairs of legs, they have few points in common. They are now placed in a distinct group of their own. Many species live in damp places on land, but a few are aquatic. They are quite common creatures, but are usually overlooked, both on account of their small size (they are never more than 1 mm long and are usually much smaller) and their cryptic mode of life.

In shape they do certainly resemble tiny bears with eight stumpy legs, bearing hooks at their extremities. The body, which consists of a head and four indefinite body segments, is translucent and almost oval in shape, but it is very flexible and can be contracted and shortened until it becomes nearly round.

The head is broad and carries sucking mouth-parts at its tip, from which food passes through a pharynx to a stomach. A body-cavity containing blood surrounds the internal organs. Two small eye-spots are present on the head.

Fig. 81   Tardigrade (× 150)

Tardigrades have no specialized breathing-organs, neither have they a heart or blood-vessels, but they do possess a simple brain, nerves and a well-developed muscular system, so that they are comparatively complex creatures.

Tardigrades are found most easily among the moss or sediment in roof-gutters, and they readily withstand the frequent drying-up and extremes of temperature which are inevitable in such situations.

At a time of drought, they contract into a barrel-shape, and have been known to remain thus in a state of suspended animation for six years. The return of moist conditions awakens them to activity again, if their exceedingly slow movements can be termed activity.

For food they suck the cell contents of plants, after piercing the cell-wall with the two sharp stylets of their mouth-parts, but it is believed that they also take animal matter such as the smaller rotifers.

The sexes are separate, and the females are both larger and more abundant than the males. They lay their eggs either singly or in groups, and sometimes these are found in the moulted cuticle of the female. The larva is a miniature of the adult, but with four distinct segments.

There are twenty genera, containing nearly three hundred species. Little is known of their distribution in the British Isles and they offer a rewarding group for study by naturalists and microscopists. The best way to obtain tardigrades for examination is to soak moss or lichen for some hours in a bowl of water and then squeeze it vigorously while still in the water. After removal of the moss or lichen the water is allowed to settle and then the liquid is decanted off to leave a sediment at the bottom just covered with water. Examination of this residue in a small saucer or petri dish under a low power of the microscope, preferably under dark-ground illumination, should reveal a large number of tardigrades, as well as rotifers and other organisms. The specimens needed for study can be pipetted to a watch glass for closer examination under a higher power of the microscope.

Members of the family Scutechiniscidae can be distinguished by their reddish colouring. The common species, *Milnesium tardigradum*, of the family Arctiscidae, is fish-like in appearance. The seventeen species of *Macrobiotus* (Family Macrobiotidae) lay eggs with characteristic decorations which assist in identification.

I7: THE VERTEBRATES

Although this book is concerned primarily with the invertebrates, or animals without backbones, no review of freshwater life would be complete without at least a brief mention of the vertebrates. The chapter will serve also, perhaps, to place in its true perspective the phylum of which we ourselves claim to be the highest members, and will indicate what a small part of the Animal Kingdom the backboned animals form.

The phylum **CHORDATA** includes the following six groups which are of interest, directly or indirectly, to the freshwater biologist:

CYCLOSTOMATA (Gr. *kyklos*, a circle; *stomat*, a mouth) — The cyclostomes or round-mouths: lampreys

PISCES (L. *piscis*, a fish) — The fishes

AMPHIBIA (Gr. *amphibios*, leading a double life, i.e. on land and in water) — The amphibians, including the newts, frogs and toads

REPTILIA (L. *repto*, I crawl) — The reptiles—lizards and snakes

AVES (L. *avis*, a bird) — The birds

MAMMALIA (L. *mamma*, breast) — The mammals, creatures that suckle their young

No reptiles, birds or mammals are true inhabitants of fresh water in Britain, but as the visits of some of them to water are not without effect on the aquatic creatures, brief mention of these must be included.

The vertebrates are important in the economy of fresh water in two ways: most of them are predators, and therefore have a considerable effect in reducing the numbers of the other creatures in the water; the decomposition of their bodies after death, and the absorption by the water of their waste products, gaseous and solid, when they are alive, must add considerable amounts of nutrient substances, an effect more important, perhaps, in smaller bodies of still water than in big lakes or rivers. The large size of the vertebrates, in comparison with the majority of the invertebrates, compensates in both these functions for their relative paucity of numbers.

The Cyclostomes

The cyclostomes although true vertebrates do not, in fact, possess a backbone made up of disc-like segments or vertebrae. In these creatures, the gristly unsegmented supporting rod or *notochord* (Gr. *notos*, the back; *chorde*, a string) persists throughout life, and does not give way to the jointed backbone as it does in more highly evolved members of the phylum *Chordata*. The cyclostomes are probably the remnants of a stock that diverged from the main trunk of the evolutionary tree. Apart from the lack of a segmented backbone, they differ from the fishes, with a few of which they might easily on a casual examination be confused, by the absence of jaws, limbs and scales, and in breathing through gill-pockets.

Two species of cyclostomes are found in British fresh water: the lampern, or river lamprey, and the brook lamprey.

The river lamprey, *Lampetra fluviatilis*, resembles an eel in appearance and is, no doubt, often mistaken for this fish. It is uniformly greenish-brown in colour, and large specimens may reach a length of 45 cm. The circular mouth of the lamprey is surrounded by a sucker armed with teeth. Inside the mouth is a tongue,

Fig. 82   Lamprey

nasal opening

gill clefts

also bearing teeth, and this can be brought forward to rasp off the flesh of any prey unfortunate enough to be held by the sucker.

Most river lampreys, in spite of their name, spend much of their life in the sea, but they return to fresh water in autumn ready for spawning in the following spring. On their return to the river they cease to feed, and the alimentary canal degenerates and atrophies. The females lay up to 40,000 eggs in a nest made among pebbles well upstream and then die.

The young of lampreys differ so markedly from the adults that they were formerly considered a separate species, *Ammocoetes branchialis*, and they are still referred to as ammocoete larvae. Other names are prides and niners, from the supposed presence of nine eyes. The larval stage, however, has no eyes exposed, and the structures which make up the 'nine' are the seven incipient gill-pouches, the single nostril on the top of the head and the area where the eye will eventually appear. The mouth of the larvae is more horseshoe-shaped than circular, and instead of teeth it has many branching tentacles inside and out. Organic particles from the mud are drawn into the mouth and entangled in mucus before being swallowed as food. The larvae lie buried in the substratum for most of their life.

Their breathing, too, is different from that of the adults. Whereas the latter take in water through the gill-clefts for this purpose, the larvae take it in through the mouth and pass it out through the gill-clefts. The larval stage may occupy five years, by which time they are 15–20 cm long and, in the adult form, they then migrate downstream.

The brook lamprey, *Lampretra planeri*, is much smaller than the river lamprey, rarely exceeding 18 cm in length. It is found in clear, rapid streams and small rivers, spending most of the daylight hours in a burrow, coming out only at night. Breeding takes place in May or June, and several females may use the same spawning site. The larvae feed on small invertebrates and detritus, but the alimentary canal of the adults is atrophied and they do not feed.

Lampreys are by some considered good eating, but with the example in mind of Henry I, who is said to have died through eating too many of them, most people will consider it advisable to leave alone this rather indigestible 'fish'.

The Fish

Freshwater fish do not form a separate division of Pisces, but representatives of many families, mainly marine, are found in rivers, streams, lakes and ponds. Some seem to have severed all connection with the sea at an early stage of their evolution, others more recently. A few fish still migrate to and fro between the sea and fresh water, either seasonally or at one definite stage of their lives, as is the case with the common eel, *Anguilla anguilla*.

Where the eels of our ponds, ditches, rivers and lakes bred was for long a complete mystery, for no one had ever seen the eggs or very small young. On the other hand it was known that the normally yellowish adult eels, when they reached a length of about 30 cm, became silvery in appearance and made for the sea as if to breed. By 1893 it was realized that the flat, leaf-shaped, transparent 'fish' to which the name *Leptocephalus brevirostris* had been given were, in fact, larval eels. Painstaking research by a Danish oceanographer, Dr E. J. Schmidt, who took sample catches across the Atlantic Ocean, showed that these larvae decreased in size and increased in abundance as he went westwards, until he located an area in the Sargasso Sea, between Bermuda and the Leeward Islands, where they were smallest—about 5–15 mm long. In 1922 Schmidt published an account summarizing the breeding of eels and showing that the larvae take three years to travel the two to three thousand miles to the rivers of Europe, making use of what used to be called the Gulf Stream, but is now known as the North Atlantic Drift.

The breeding-grounds of the slightly different American eel, *A. rostrata*, overlap those of the European species, but are centred to the west and south of its breeding area. The larvae of the two species intermingle in mid-Atlantic, but those destined for the American coasts separate and reach their destination in about a year.

When they are nearing the end of their journey, the larvae are about 75 mm long and they then become rounder, narrower and shorter, although still trans-lucent. These 'glass eels' turn to 'elvers', and in April or May the active little creatures make their way up rivers, often in huge numbers, with the head of one almost touching the tail of the one in front, a solid procession which country people call the 'eel-fare'. Obstacles in their route are overcome with persistence, and it is at this time that they will even cross land to reach suitable stretches of water. They remain in freshwater habitats for from five to twenty years before turning into silver eels and setting off once again for the sea.

It has been assumed that both European and American eels make the journey to the Sargasso Sea to breed and then die, but in 1959 a British zoologist, Dr D. W. Tucker, put forward a controversial theory that adult European eels do not breed, and die without crossing the Atlantic, the stocks of European eels being the offspring of the American eels. He explained the morphological differences between the two kinds by differences in current and temperature in the breeding area. There has been considerable discussion on this novel hypo-thesis and it has not been generally accepted. Possibly new discoveries still to come may solve eventually the age-old mystery of the eel.

Whereas the eel migrates to the sea to breed, the salmon does the exact opposite. Hatched in fresh water, it makes its way in due course to the sea, but later returns to the river to breed.

To deal adequately with all the fish that are found in fresh water would need a volume larger than the present one, and the reader who wishes to find out more details is recommended to consult the books mentioned in the bibliography on p. 272. Here little more can be done than mention a few representative fish from each family.

SALMONIDAE   Included in this family are perhaps the best-known freshwater fish and the most prized, not only because of their edible qualities, but also on account of their beauty.

As we have seen, the salmon, *Salmo salar*, is a freshwater fish for only a small part of its life, yet commercially it is the most important of all freshwater fishes. A full-grown male may measure 150 cm in length and weigh 38·5 kg (85 lb). In its adult stage it probably has little effect on the other organisms in the river, for it feeds little if at all in fresh water. After spawning, both male and female, known at this stage as 'kelts', are feeble and thin from starvation and the rigours of the breeding period. Once again they make their way back to the sea, but many, particularly the males, perish on the journey.

The eggs of the salmon are laid in a depression—the 'redd'—made in the gravel of the river bed by the female. Here the male fish fertilizes the eggs by shedding his milt over them. The newly hatched salmon, or 'alevin', with its egg-yolk attached to the underside of its body, spends its early days in the gravel of the river bed, but after about a month the food in the yolk-sac is all used up and the little fish, now called a 'fry', seeks out its food. At first microscopical planktonic animals, and small crustaceans, are taken. When about eighteen months old the young salmon is some 15 cm long and the transverse dark-blue bands, or 'parr-markings', along its body have disappeared and the now silvery 'smolt' makes its way to the sea in search of richer feeding than the river can offer.

The length of time the young fish remain in fresh water varies, however, and some may not migrate until their fourth summer. Similarly, the time spent in the sea varies. Some return to breed after only a year; these are called 'grilse'. Others may be four years or more before they return to breed in the river. Little is known of the salmon's life in the sea, but in 1965 extensive feeding grounds were discovered off the west coast of Greenland. Inevitably, this led to large scale exploitation in the form of a large gill-net fishery, which has given rise to concern about the survival of the Atlantic salmon, especially as this coincided with a serious outbreak of 'salmon disease', ulcerative dermal necrosis (U.D.N.), which attacked several species of salmonid fishes for some years and seemed to involve several pathogens including a virus, slime bacteria and the fungus *Saprolegnia*.

Experiments carried out by marking fish have shown that salmon tend to return to spawn in the actual river in which they themselves passed their early days. Salmon enter river estuaries all through the year, but the main 'running', i.e. the migration up the rivers to spawn, takes place between spring and autumn, and at this time, driven by an irresistible urge to reach the shallow waters in which they will breed during the autumn or winter, they surmount all obstacles in their route, leaping over weirs and up waterfalls. On many salmon streams man, activated not solely by altruistic motives, has erected salmon-ladders to facilitate the negotiation of such obstacles.

The trout, *Salmo trutta* (and it will be assumed that there is but one species and the sea trout—which, like the salmon, migrates between the sea and fresh water—lake trout, brown trout and the rest are merely varieties of *Salmo trutta*), is another handsome fish, a delight to the eye in colouring and form, as well as a treat to the palate. Little wonder that it has been introduced to countries all over the world as the sporting fish *par excellence*. Although giants of over 100 cm in length and many kilogrammes in weight do occur, the trout from the small, swift streams in which they seem most at home are more in the region of 170–225 g and measure only just over 20 cm when fully grown. Running water is, however, not an essential, and trout thrive in lakes and even in well-oxygenated ponds. The colouring varies greatly among the many varieties, but an unfailing characteristic is the presence of dark spotting over the body, a feature which makes a trout extremely difficult to see as it rests over the bed of a stream dappled with sunlight. Trout are exclusively carnivorous, and stomach examinations have shown that the smaller specimens subsist largely on insects (winged and aquatic), crustaceans and snails. Larger specimens take toll of other fish, such as minnows, small eels and bullheads, and may even turn cannibal. The catholicity of their diet has been an important factor in enabling trout to be introduced into fresh water in many parts of the world. The breeding of the trout is similar to that of salmon. Eggs are laid about November, and the alevins hatch in late February or early March. (Plate 53, top)

The char, *Salvelinus alpinus*, is similar in appearance to the trout, the spotting, however, being lighter in colour, and the leading edges of the pectoral, pelvic and anal fins are white. Local varieties of this fish are found in deep lakes in various parts of the country. The male char in the breeding season is a most handsome fish with vivid reddish underparts. The eggs of fish spawning in autumn are laid in the shallow waters of the lake or in the small streams running into it, but those that spawn in spring do so in deeper water.

THYMALLIDAE   The grayling, *Thymallus thymallus*, although perhaps not quite so handsome as the other fish which have been mentioned, is nevertheless a very beautiful fish, and its flesh is considered by some to be more tasty than the trout. The common name refers to its greyish colour, and its scientific names to the aroma of thyme which is detectable in the flesh of the freshly caught fish. The grayling is found mainly in swift, stony streams and grows to a length of 50 cm. Spawning takes place in spring. The fish is carnivorous and feeds on freshwater shrimps, snails, caddis larvae and other invertebrates that live on the bottom.

ESOCIDAE   The pike, *Esox lucius*, our only representative of this family, is the most ferocious of the freshwater fish, and attracts local names such as 'water-wolf' and 'king of the lake'. Not only fish, but water birds, water voles and frogs are caught and devoured by this savage creature. Lakes and slow streams are the haunts of pike, the eggs being laid in spring among the weeds.

Although the activities of the pike seem at first sight to be all on the debit side, it must be remembered that it serves a useful purpose in keeping in check the smaller fish and thus indirectly increasing the number of more minute creatures which would otherwise be devoured.

CYPRINIDAE   To the carp family belong the majority of the so-called coarse fish.

The common carp itself, *Cyprinus carpio*, although found in many parts of the country, is probably not a native fish but was introduced long ago by the monks as a food fish. On the Continent it is still the most important freshwater fish to be reared artifically for food. There are a number of varieties, such as the mirror carp which has patches on the body devoid of scales, and the leather carp which is almost scaleless. As a species it prefers warmer water and still, weedy water. It is a bottom feeder, taking both animal and plant food.

The gudgeon, *Gobio gobio*, is common in many parts of the country, mainly in streams. It is slender in shape, and is readily identified by the *two* barbels hanging down on either side of the mouth. It is a bottom feeder and takes both plant and animal material. Spawning takes place in May and June, the eggs being attached to stones, plants or the bottom gravel.

The tench, *Tinca tinca*, is found mainly in still or slow-moving water and prefers weedy conditions. It feeds on vegetable material and on small bottom-living animals. The tench is a favourite fish on Continental fish farms because of its tenaciousness to life under adverse conditions.

The minnow, *Phoxinus phoxinus*, is a lively little inhabitant of clear streams. In the breeding season in early summer, the colour of the male, normally dull greenish-brown, becomes more intense, and reddish patches occur near the mouth and on the gills. As an added embellishment, and in common with other members of this family, white tubercles appear on the back and head. Spawning takes place from March to June, the eggs being attached to stones. Minnows feed mainly on insect larvae and small crustaceans, but take also some plant food.

The chub, *Leuciscus cephalus*, is widely distributed in England in still and slow-moving waters. Although the average size is between 30 and 50 cm, they may reach 60 cm. The sticky eggs are fastened to stones and plants in May or June. Chub feed on small invertebrates and plants, but when they are larger they may take small fish, frogs and crayfish.

The dace, *L. leuciscus*, is a silvery and very lively little fish. Although mainly a fish of streams, it is also found in lakes. Spawning takes place from March to May. Dace feed on a wide range of animal food and only rarely on plants.

The roach, *Rutilis rutilis*, is a very common fish in still and slow-moving waters, but it is not always easy to distinguish it from the rudd, *Scardinius erythrophthalmus*, found in similar places. Both have reddish pectoral, pelvic and anal fins, but the sides of the body, which are silvery in the roach, tend to be more golden in the rudd. The eggs of both are attached to water-plants in May or early June, in the case of the roach, somewhat later by the rudd. Both are omnivorous.

The bream, *Abramis brama*, a larger fish which frequently reaches a weight of 3 kg (6½ lb) or more, is very deep in body and thus easily recognized. Spawning takes place from May to July, the eggs being deposited on plants. Small crustaceans and insect larvae provide the main food, but larger specimens take snails.

The last member of this family to be considered is the bleak, *Alburnus alburnus*, a lively little fish whose abundant energy often causes it to leap out of the water to capture insects flying over the surface. Its eggs are attached to stones in early summer. In France this fish is reared especially for its scales, from which a substance is obtained for use in the manufacture of artificial pearls.

COBITIDAE. The most familiar member of this family is the stone loach,

Noemacheilus barbatulus, which can be found hiding among the stones of rapid streams in most parts of the country. It is about 10–13 cm long and a distinguishing feature is the fringe of barbels around the mouth. Stone loaches feed on any small animals living on the bottom.

ANGUILLIDAE   The common eel, *Anguilla anguilla,* has already been mentioned. It is a common inhabitant of streams and rivers in most parts of the country, and relished as one of the most nutritious of freshwater fishes. It feeds on all kinds of aquatic animals and even small rodents and water birds.

PERCIDAE   One of the most handsome of our freshwater fish is the perch, *Perca fluviatilis,* which is common in streams, lakes and ponds. The spiny dorsal fins, reddish in colour, are a distinguishing feature. The eggs are laid in sticky lace-like bands which can usually be seen floating on the water in early summer attached to water-plants. Although specimens up to 2·25 kg in weight may be taken, much smaller ones are usually found. Small fish feed on water fleas found among plants in shallow water. Later they graduate to insect larvae and freshwater shrimps, while large specimens take other fish.

During the 1939–45 war, in an endeavour to reduce the multitudes of small perch in Lake Windermere, the Freshwater Biological Association trapped them in large numbers, and they were canned by a commercial firm and marketed under the name of 'Perchines'. Over a period of five years some eighty tonnes of the fish were trapped. The public did not apparently accept this new addition to their menu with any great enthusiasm, in spite of the stringencies of the time, but the experiment did give some indication of the vast number of perch in the lake.

The pope or ruffe, *Gymnocephalus cernua,* is a smaller and rarer relative of the perch.

COTTIDAE   The bullhead or miller's thumb, *Cottus gobio,* is plentiful in streams and rivers, its flattened body, about 8 cm long, being well adapted to a sedentary life in running water. Most of its time is spent hidden between or under stones waiting until some insect larva or freshwater shrimp comes within sight, when it darts out to snap it up. Its spines no doubt protect it from molestation by larger fish or other creatures. In March or April, a depression is scooped out under a stone and the eggs laid there.

GASTEROSTEIDAE   The three-spined stickleback, *Gasterosteus aculeatus,* is perhaps the first freshwater fish with which most of us make contact, for it is the well-known 'tiddler' which, in great numbers, comes to an untimely end in the jam-jar aquaria of all budding naturalists. (Plate 56)

It is a handsome, lively little fish, about 5 cm long, and the male when resplendent in the breeding colouring is a jewel indeed. There can be few stretches of still or slow-moving water where sticklebacks cannot be found, and they even make their way into the brackish water of river-mouths. At breeding time, in spring, the male makes a nest from pieces of water plant and detritus glued together with a substance secreted from his kidneys. Several females may lay their eggs in the nest, and until the young are able to fend for themselves the male stands guard over the nest, chasing away intruders and even catching in his mouth recalcitrant youngsters which are straying from the nest before they are ready, and spitting them back.

Sticklebacks feed almost exclusively on living animal matter, a characteristic often, with fatal results, overlooked by those who try to keep these interesting little fish in aquaria to study their nest-building habits. The ten-spined stickle-back, *Pungitius pungitius*, is not so common and is less widespread in distribution. Both species are sometimes found in brackish water.

To conclude this section, it may perhaps be of interest to give in tabular form some results of an investigation carried out at the Freshwater Biological Association on the food of freshwater fish. The figures are the occurrences of the different foods expressed as percentages of the total number of occurrences.

	FISH	MOLLUSCS	INSECTS	CRUSTA-CEANS	PLANTS AND FILA-MENTOUS ALGAE	MICRO-SCOPICAL ALGAE INCLUDING DIATOMS
Pike	88	—	8	4	—	—
Eel	18	7	29	36	10	—
Gudgeon	—	16	56	19	9	—
Dace	—	16	32	22	26	2
Roach	—	8	11	14	55	12
Rudd	1	—	41	9	48	1
Bream	—	4	20	51	23	2
Perch	4	—	54	41	1	—
Three-spined Stickleback	—	3	60	29	4	4

The Amphibians

The amphibians have evolved from fish-like ancestors and, although they have colonized dry land, they still show unmistakable traces of their fish ancestry. They must, for instance, return to the water to breed, and most of them have an aquatic larva which breathes by means of gills. The adults, however, have lungs and the nostrils, through which the air is taken in, open into the mouth.

Our amphibians belong to two orders:

CAUDATA (L. *cauda*, tail): The newts.

SALIENTIA (L. *salio*, I leap): The frogs and toads.

Britain is not rich in amphibians, and only eight species are found here, two of which are not really natives but have been introduced in recent times.

The Newts

Three species of newts occur in Britain. The common or smooth newt, *Triturus vulgarus;* the great crested newt, *Triturus cristatus*; and the palmate newt, *Triturus helveticus.*

The smooth newt is quite the commonest of our newts, and is found all over England, in Ireland (where it is the only species), and in parts of Wales and Scotland.

The female is of a dull brownish colour, but the male is more brightly coloured, with heavy black spotting over the body and a reddish belly. In the breeding season the colours of the male become even more vivid, and the serrated crest, which at other times is no more than a ridge along the back, becomes very pronounced. (Plates 54, 3; 56)

During most of the year, newts live on land and are rarely seen, for they hide under stones or other objects during the day and feed by night. In winter they hibernate in nooks and crannies. It is only in spring, when they take to the water, that they are much in evidence.

After a short period of courtship the male deposits a transparent conical sperm-capsule in the water, and the female, grasping this with her vent, releases the sperms into her body and so fertilizes the eggs before laying them. She attaches them singly to water-plants, and often, when the egg is laid on a leaf, part of the leaf will be bent over to shield it. Soon after egg-laying is completed, the adults leave the water.

The larvae, which hatch out of the eggs in about a fortnight, are delicate little creatures with three pairs of external feathery gills on the side of the head. The front pair of legs appear first and the hind legs appear later. By mid-August the little newts are ready to leave the water, having by that time lost their external gills and being able to breathe atmospheric air, but some remain in the water until the following year. (Plate 54, 2)

The great crested newt is a much larger creature than either of the other two species—about 15 cm in length against the 10 cm of the smooth newt and the 7·5 cm of the palmate newt.

Both the male and the female are darker than in the other species, and the skin is covered with small tubercles that secrete a substance which is distasteful to other creatures; doubtless this secretion serves as a means of protection. The life-history is similar to that of the smooth newt, but adult great crested newts spend more time in the water.

The palmate newt is probably commoner than is popularly supposed, and is no doubt frequently confused with the smooth newt. Although the females of the two species are hard to distinguish, it is not difficult to recognize the male palmate newt, particularly in the breeding season, by his possession of a thread-like filament at the end of the tail and by the webbing of the hind feet. After the breeding season the tail filament is reduced and the webbing of the feet becomes little more than a fringe of skin on the toes.

The food of newts when on land consists of insects, worms and the like. In the water they and their larvae feed on any small animals they can catch. In my younger days a favourite method of capturing adult newts during the breeding season was to tie a live worm to the end of a line of cotton and, a few centimetres away, a match-stick to act as a 'float'. If this bait was dangled at evening in a pond from the end of an improvised fishing-rod, it was not long before newts would be nibbling, and as soon as the match-stick disappeared the newt could be hauled in, firmly gripping the worm in its jaws.

The Toads and Frogs

Our toads can readily be distinguished from the frogs by their warty skin, and also by their shorter legs which enable them to move along by crawling, whereas the frogs progress by hops.

The common toad, *Bufo bufo*, strays far from water during the summer and autumn, and hibernates in deep holes from October to March. Early in spring the males and females make their way to permanent stretches of water, and after a protracted courtship the eggs are laid by the female in long ropes of gelatinous material, the male depositing his sperm over them as they are laid. These egg-ropes may be 3·5 metres or more in length, and sometimes contain as many as 7,000 eggs. The tadpoles hatch in about ten days and complete their metamorphosis in about three months. When first hatched they have external gills, but these are absorbed at the end of a month and internal gills are then used for breathing. The hind legs appear at about seven weeks, but the fore legs, although developed, do not burst out of the skin until a few days before the tadpoles leave the water. At the same time the tail is absorbed, the gills atrophy and the creatures begin to use their now developed lungs by rising to the surface of the water for air. (Plate 53, middle)

On leaving the water the tiny toads are only a little over 12 mm in length, and it is not until they are about five years old that they are mature and ready to breed. (Plate 57, 5)

The tadpoles of both toads and frogs are, of course, one of the most important sources of food for the larger carnivorous freshwater creatures, and incidentally for each other, for cannibalism is the order of the day if other animal food is scarce.

The natterjack toad, *Bufo calamita*, although rarer than the common species, is found in considerable numbers in certain localities, usually in sandy districts. It is more agile than its relative and can run rapidly. The body colour is greener, and the yellow line down the middle of the back instantly identifies it. When alarmed, the natterjack becomes most vividly coloured, the background colour becoming lighter and the blotches more pronounced.

During the day toads live in burrows which they excavate in the sand, but occasionally they are seen running about in full sunshine. The noise made by the males during the mating period, from the end of April to early June, is quite incredible. It is not so much a croak as a continuous gurgle, and can be heard over 1·5 km away on a still night.

The tadpoles of the natterjack develop more quickly than those of the common toad, leaving the water in about six weeks. This is possibly an adaptation against the drying-up of the temporary pools which the natterjack is, by the nature of its habitat, compelled to use. (Plate 57, 4)

The common frog, *Rana temporaria*, is well known to everyone, and its tadpoles rival the stickleback as the first aquatic creature of which we become aware in childhood. Full-grown common frogs may be 75 mm or more in length, and the ground colour varies enormously through a range of greens, browns and yellows. The great masses of spawn can be found in nearly every patch of water, temporary or permanent, in early spring. The tadpoles pass through similar phases to those of the toads, development being completed in about twelve weeks. (Plates 53, bottom; 54, 1)

The edible frog, *Rana esculenta*, is not a native of this country, but repeated attempts to introduce it have been made. The Fens, Surrey, Hampshire, Oxfordshire, Bedfordshire, Middlesex and Kent are counties into which it has been introduced at some time or other. It is more aquatic in its habits than the common frog, and if alarmed when on land immediately makes for the water into which it dives with a loud 'plop'. Although variable in colouring, it is in general greener than the common species and slightly larger. The hind legs of the edible frog provide the delicacy which is so greatly relished in France and other parts of the Continent. (Plate 57, 2)

Another comparatively recent addition to our fauna is the marsh frog, *Rana ridibunda*, a large creature the females of which may measure up to 12·5 cm in length, the males being smaller. The body is greenish-brown in colour, usually spotted with black. On account of its size, the marsh frog is preferred to the common frog for experimental work, and was frequently imported for this purpose from the Continent before 1939. The present wild stock in this country is said to have originated from twelve specimens which were brought from Hungary in 1935, and released in a garden at Stone-in-Oxney, in Kent. These apparently found a suitable habitat in the many dykes which intersect this part of the country. Romney Marsh and adjoining areas are now districts where the marsh frog is established, and attempts have been made to introduce the species in other parts of the country. It is to be hoped that such attempts will not succeed and will not be repeated, for the noise of a chorus of males at breeding time is said to be sufficiently loud to keep a whole village awake! (Plate 57, 3)

The Reptiles

The British reptile that is most likely to be encountered in water is the grass or ringed snake, *Natrix natrix*. It is sometimes seen swimming gracefully, with a side-to-side movement of the body, the head being held a few centimetres above the water. Frogs are its staple diet, but fish, newts, insects and tadpoles are also eaten. The prey, if caught in the water, is brought on to land and there swallowed. The viper, or adder, *Vipera berus*, usually frequents dry places but does take to the water sometimes and swims well.

The Birds

There are few species of birds, other than pelagic sea-birds, that do not visit stretches of fresh water regularly or periodically, if only for the purpose of drinking or bathing. But there are some which we automatically associate in our minds with water. The brilliantly-coloured kingfisher, perched on its favourite branch surveying the water below and ready to plunge when it sees its prey; the dainty little dipper flitting from stone to stone in a hill-stream, or perhaps swimming below the surface in search of the insects on which it feeds; the stately swan; the ducks; the moorhen and coot; all these and many more play an immensely important part in the economy of fresh water, quite apart from the predatory function which immediately springs to mind. Their freedom to move from place to place, and the fact that many of them do in the course of their migrations traverse vast distances, make them efficient vehicles for the conveyance of organisms, both plant and animal, from place to place, and thus they have a profound effect on the distribution of aquatic life.

Darwin first pointed out in his *Origin of Species* how readily wading birds, many of which undertake particularly long migrations, might convey seeds in mud adhering to their feet. But in addition to seeds and vegetative fragments of plants, it is now clear that much animal life can be similarly transported, including molluscs, crustaceans, water mites, rotifers, sponge gemmules and the statoblasts of bryozoans. So far as plant life is concerned, there is some evidence that seeds which have passed through the digestive tracts of birds germinate more readily than those which have not done so. In one instance of nutlets of the pondweed, *Potamogeton natans*, eaten and passed by a domestic duck, sixty per cent germinated the following spring, whereas only one per cent of those left over in the vessel from which the duck had fed did so. When we remember the number of seeds eaten by birds—828 seeds and fruits including those of bur-reed, *Sparganium*, and *Potamogeton* were found on dissecting thirteen wild duck purchased in a London market—we can have little doubt of the importance of birds in the distribution of plants.

The Mammals

Only three mammals—the otter, the water vole and the water shrew—merit consideration as waterside creatures, although, as with the birds, a number of others visit the water and no doubt have some influence on the transport of plants and animals to adjacent stretches of water.

In the past, horses and cattle drinking at both village and field ponds, played an important part, by their trampling of emergent vegetation, in keeping ponds 'open'. Modern farming practices, and especially the universal supply of piped water, restricted this use of ponds with the result that many have become silted up by the invasion of plants; some have even disappeared or been filled up to become part of the surrounding land.

The otter, *Lutra lutra*, although rarely seen since it is a nocturnal animal and a shy and wary one at that, is commoner than is popularly believed where there are waters suitable for it. Some years ago when collecting information on the occurrence of mammals in a comparatively built-up area, I was surprised at the number of records, which diligent inquiries revealed, of dead otters being found, usually electrocuted on a railway line. A male (dog) otter weighs about 10–12 kg and measures from nose to the tip of the tail 106–120 cm. The females (bitches) are smaller. The 'holts' in which the creatures rest all day are usually in the banks of streams, a cavity under the roots of trees being frequently used. At sunset they come out to hunt, entering the water silently and swimming at the surface. They capture their prey in the water, but eat it on land. Fish of all kinds and crayfish, or, failing these, frogs, birds and almost any living creature, are taken as food. Although the angler deplores the number of salmon and trout consumed, he should remember that the otter serves a useful purpose as well in ridding the water of ailing and cannibal fish, and also of eels which, being carnivorous and also greedy, are one of the worst pests in a river. It would be a sad day if the otter's whistle were silenced for ever.

There can be few people who, when frequenting the waterside, have not heard the 'plop' of a water vole diving into the water. The water vole, *Arvicola terrestris*, or, as it is often unkindly called, the 'water rat', is common throughout Britain, frequenting sluggish streams and drainage dykes, and doing a certain amount of damage to the latter by tunnelling. They feed by day and are almost, if not exclusively, vegetarians, eating the stems and roots of marginal plants. The water vole is quite a large animal, measuring from 15–22 cm in length, excluding the tail. The snout is blunter than that of a rat, and the ears are lower on the head and not easily visible. The creature is well adapted to an aquatic life by having an oily skin, and a modification of the ears to prevent water from getting in. The young are born in early summer in a nest of partly chewed fragments of water-plants, located usually in a burrow.

The water shrew, *Neomys fodiens*, the third of our waterside mammals, is also the smallest, being only about 7·5–10 cm in length, excluding the tail, and weighs only about 15 g. It is of the typical shrew form with a long pointed snout, but is adapted to an aquatic life by having the feet and tail fringed with hairs to aid in swimming, and by possessing the ability to close its ears when under water. Shrews hunt both by day and night, and aquatic insects, fish spawn and fry, as well as slugs, worms and insects caught on land, form their diet. Nests of grass leaves are made in branched tunnels in the banks of clear, slow streams, and there the young are born about May.

This completes our survey of aquatic and waterside vertebrates, but there is one animal that we have not yet considered in this chapter, although it has, perhaps, more influence on the economy of fresh water than all the others put together. Man, in many ways—by his agriculture, his commerce, his search for water supplies, his sewage disposal, his leisure pastimes—has profoundly changed, and is still altering, the character of much of the fresh water in this crowded little country, and with it the nature of the aquatic life. This subject is discussed in more detail on pages 256 to 258.

I8: STUDYING FRESHWATER LIFE

The attitude to natural history collecting has changed radically in recent years. The ever-increasing loss of natural habitats resulting from urban and industrial development and intensive agriculture and forestry has made it necessary to adopt a more responsible approach to natural history studies if the flora and fauna are to be conserved. In no branch of natural history is this more necessary than in freshwater studies for, concurrently with the rapid disappearance or deterioration of smaller bodies of water, greater pressure has arisen on those that remain. Colleges, schools and the many outdoor pursuits centres of local authorities invariably include these studies and parties make repeated visits to areas of rich natural history interest. Collecting, which only a few years ago would have had a trivial effect on the freshwater fauna and flora, can now, as a result of these changes, affect the very survival of species if continued without restraint.

It seems necessary, therefore, before discussing the subject of this chapter in detail to set out a few rules based on the separate Conservation Codes drawn up by the Joint Committee for the Conservation of British Insects and the Council for Environmental Education, to whose valuable initiative all British naturalists should be grateful:

1 If possible, examine all larger specimens at the waterside and return them as soon as possible to the water, making field notes at the time.

2 If some specimens are essential for further study, take no more than are necessary, keep them in cool, shady conditions, use proper containers to take them home and return them, if possible, to their original habitat, as soon as they have been studied. If conducting a party, appoint one person to be responsible for specimens.

3 If stones or logs are turned over in the water to look for specimens, remember to turn them back again or the animals which live underneath may die.

4 Water-plants which have been taken out so that they can be examined for specimens must be returned to the water.

5 Avoid trampling the ground along stream and pond margins, and leave the habitat as you found it.

6 Always ask permission from landowners or occupiers to study on private land.

One of the greatest delights of the study of freshwater life is that it can be carried out almost anywhere, and in this respect it offers considerable advantages over most other branches of natural history. There are few places in our country where some stretch of fresh water cannot be found; even in big cities there are usually boating-lakes or other ornamental waters which the enthusiastic naturalist can study, and if all else fails, the rain-tubs or bird-baths in his garden, or those of his friends, will provide material in abundance. The equipment with which to pursue the pastime need be only of the simplest; if one or two small specimen tubes and a few polythene bags are slipped into the pocket when setting out for a country walk, enough material for hours of study with the microscope can be obtained by filling them with samples of the water, or a few fragments of water-plants, from any ponds or streams encountered.

For the keen student who wishes to pursue his interest more seriously, how-ever, there are certain basic items of equipment which will simplify his work and enlarge its scope. Since this book was first written it has, unfortunately, become impossible to buy from commercial sources many of the items formerly available

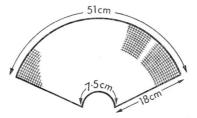

Fig. 83   Pattern for small plankton-net

for the study of freshwater life. Most of those described below, however, are fairly simple to make or improvise by anyone who has access to a good workshop.

First of all, some form of pond-net is almost essential. Probably the type of most general utility is that known as a *plankton-net*. (Plates **58, 1, 2; 62**) The bag part is made of fine-mesh material, and carries at its base a tube or bottle into which the organisms descend when the net is lifted out of the water, so that they are readily seen and, if desired, examined with a lens. Although such nets were, of course, primarily intended for obtaining the tiny floating planktonic animals and plants, they can, if used with care, be employed quite successfully as all-purpose pond-nets, and it is a great convenience to see what has been caught without having to empty the net. Such a net can readily be made by anyone able either to stitch two pieces of cloth together or to persuade some female relative to do so! The shape of material required is given here (Fig. 83).

A suitable frame to which to attach the bag can be made of a 10 mm diameter brass rod bent to the shape desired, and provided with a threaded limb which will screw on to the end of a suitable stick. Any metal-worker could make such a frame for those unable to do so themselves, or, failing that, the complete net may be bought from dealers in natural history equipment, for these are items that are, happily, still available.

Although fine muslin will serve for the net, a much better cloth is bolting silk, which is a very strong, fine-meshed material used by millers for sifting flour. It is made in a number of grades depending on the fineness of the mesh. Coarse (twenty meshes per 25 mm) is suitable for general work; medium (sixty meshes per 25 mm) for zooplankton; and fine (180 meshes per 25 mm) for the smaller microscopical creatures and phyto-plankton. Bolting silk, which is obtainable from some textile manufacturers, is expensive, particularly in the finer grades, but if treated with care will last a long time. Nylon fabric is now used successfully for making nets and is, perhaps, easier to obtain nowadays.

The receptacle at the bottom of the net may be any small wide-mouthed bottle or tube with a lip and is tied tightly to the net. Glass tubes are easily broken, so

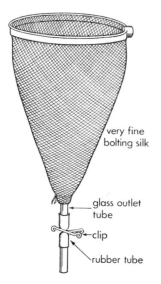

very fine bolting silk

glass outlet tube

clip

rubber tube

Fig. 84 Large plankton-net with open outlet tube

plastic tubes or bottles, such as those in which various tablets are bought, are preferable. It is an advantage if the capacity of the tube or bottle is not more than that of the type of tube in which specimens are to be carried home, so that the whole contents of the net tube can be transferred without the risk of losing part of the 'catch'.

Some workers, instead of attaching a closed bottle to the net, use an open glass tube, sealing the bottom end by slipping over it a short piece of rubber tubing bearing a spring clip. When the clip is released, the contents of the tube can readily be run into a specimen tube underneath (Fig. 84).

The choice between a circular and a square type of net is largely a matter of personal preference, although the latter, which can have a sharp metal rim, is of advantage for scraping submerged tree trunks, canal walls or posts—favourite locations of a number of organisms, such as bryozoans, sponges, some diatoms, etc. (Plate 55, A)

Commercially-made nets are often provided with a standard screw-thread which fits various types of pond-sticks. Perhaps the best of these sticks is the one in sections each of which bears a female thread at one end and a male at the other, so that a stick of almost any desired length can be fitted together at the pond-side.

Those who are sensitive to the possibly uncomplimentary remarks of others (and in this category the author must, unfortunately, number himself) will prefer a walking-stick type of net-handle. One of these is easily made by removing the ferrule from an ordinary walking-stick and substituting a metal end threaded to take the collecting net. A solid ferrule, to screw over the end, will prevent the thread from becoming damaged or filled with dirt. The commercially made sticks which used to be available were hollow, and had an extension rod inside which doubled their length when used as a pond-stick.

There are a number of other accessories which can conveniently be used on a pond-stick. Thus an ordinary gauze net, without a tube, may be fitted when the nature of the specimen, the presence of stones or dense weeds in the water make a plankton net unnecessary or unwieldy. A satisfactory way of keeping leaves, sticks, etc., out of the net is to fit a strainer top (coarse muslin stretched on a wire frame) to the open end of the net. (Plate 59, 2)

Another useful accessory which can be screwed to the standard stick is a bottle-clip (Plate 55, C), which will, by adjustment of a sliding piece, grip a specimen bottle round the neck, and enable a sample of water to be collected far out from the bank or deep in the water. To cut off pieces of water-plant, a bent or folding knife is useful (Plate 55, D), and for skimming the surface of the water for desmids or diatoms, a spoon strainer can easily be made (Fig. 85).

perforated gauze

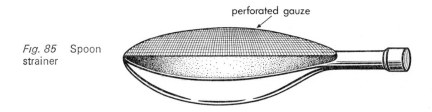

Fig. 85 Spoon strainer

For obtaining portions of water-plant far away from the bank, a drag of some kind is necessary. This is merely a series of prongs or hooks, weighted and attached to a long length of strong twine. Cast out as far as it will go and then dragged slowly back to the bank, it will entangle and uproot most of the plants over which it passes. This accessory, however, is often a most exasperating piece of equipment to use, for it seems to go out of its way to become entangled in submerged roots or other immovable objects, where it remains firmly fixed.

A simple drag can easily be made out of a short piece of lead tubing, through which are pushed several very stout wires, bent at one end into prongs, and at the other into a hook or ring for the attachment of twine. If the piping is then flattened by hammering, the wires will be securely fixed. Plastic bags are invaluable for carrying plant material home.

Small creatures from either the surface of the water or the surface of the mud can be collected by a length of rubber tubing with a bulb at one end and an open glass tube at the other. By drawing the glass end slowly along the surface, while

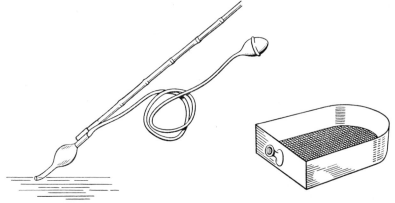

Fig. 86   Left: 'diatom bulb'. Right: 'shell scoop' which when in use is screwed on to a normal collecting stick

releasing the pressure on the rubber bulb, minute organisms such as diatoms, desmids and amoebae, will be sucked into the tube (Fig. 86). Some creatures live habitually in the mud, and for them the most convenient appliance is a scoop with a perforated bottom, which can easily be improvised by a handyman. (Plate 58, 5; Fig. 86)

When it becomes necessary to examine the bottom-living creatures over a wide area of a stretch of water, a dredge is required. The framework of this, either triangular or rectangular (Fig. 87), is solidly made, usually of iron and heavily weighted to keep it on the bottom. The net portion should be long to accommodate the large quantity of bottom deposit which it will collect in its passage through the water. Stout tow-ropes attached to the corners of the frame serve to

Fig. 87   Dredge

pull the dredge along. In ponds the dredge can be placed in the water at one end and, after the operator has walked to the other side with the tow-rope in his hand, pulled across. In lakes it is usual to take the dredge out in a boat, drop it overboard and then haul it in from the shore.

Tow-nets (Plate 62) are used to collect organisms from water levels nearer the surface. They are essentially very long plankton-nets with a tube at their apex, but with tow-ropes attached to their open end instead of a stick. They are usually towed behind a boat or from one side of the stretch of water to the other. Floats are sometimes fastened to them to keep them at the level desired.

For examining specimens at the waterside, a shallow white dish is useful. Into this each sweep of the net, or haul of the drag or dredge, can be placed. Against this white background most forms of life, even the very small, will be seen readily, and when the water and debris are eventually thrown away, small molluscs, planarian worms and other creatures which might otherwise be overlooked will be found adhering to the bottom of the dish. Enamel pie-dishes, photographic developing dishes and the like are admirable for this purpose. Specimens of the various organisms can be transferred to a tube, either by a pipette or by forceps (Plate 55, F, H), and in general each form of life should be kept to a tube of its own. This not only simplifies the sorting out of the creatures at home, but may prevent some cannibalism or murder. Insect larvae and similar larger creatures are better carried in small tins packed in damp weed or grass than in tubes of water, and anglers' small bait-tins (Plate 55, I) are excellent for the purpose as they are rustless and have a gauze-covered air-hole in the lid.

Stoppered specimen tubes in various sizes and diameters are sold by dealers for carrying specimens; for general use the most suitable size is 75 mm long by 25 mm wide.

A useful piece of equipment to have at the waterside is a glass trough (Plate 55, E). This is a very small aquarium, about 100 mm by 75 mm, but only 12 mm or so from front to back. Such a trough can easily be made from two pieces of glass. Old photographic plates, quarter-plate size, which have had the emulsion cleaned off, are excellent, with a U-shaped piece of rubber or bakelite between, the whole cemented together with any adhesive impervious to water. In these troughs small pieces of plant can be placed in water and examined closely under a good lens for the presence of organisms such as case-building rotifers, bryozoans, hydra, etc. Quite a satisfactory trough can be improvised from two small pieces of glass gripped together by spring clips, with a thick rubber band between them. (Plate 59, 1)

At the end of a pond-hunt a good deal of water will have been accumulated, and to save carrying this home and yet not lose any of the creatures, a 'concentrating bottle' is useful. This is a wide-mouthed jar in the stopper of which are two holes, one to take a small funnel and the other to take an outlet tube, which must be covered at the bottom end with gauze or bolting silk. The water from the tubes is poured down the funnel and, when the jar is full, the surplus water runs out of the outlet tube, but the organisms are prevented from getting out by the meshes of the material. The plankton-net itself can, of course, be used in a similar way and the planktonic 'brew' concentrated into the tube. (Plate 55, G)

For capturing the adult stages of insects such as caddis flies, china mark moths and the like, which are to be found among the vegetation at the edge of a pond, some kind of butterfly-net will be needed. A convenient type, illustrated in Plate 58, 3, 4, has a frame made of springy steel which enables the net to be twisted into a very small compass, so that, if necessary, it can be carried in the pocket. A slight untwisting causes the net to spring open to its full extent. A cane is inserted into the tube on the frame to serve as a handle.

Finally, to complete our list of equipment, a good pocket lens should be carried on all expeditions. One with a × 5 magnification is advisable, preferably aplanatic and, for convenience, it should be slung from the neck by a cord.

The somewhat extensive list of equipment which has been mentioned in this chapter is not by any means necessary for every visit to a pond or stream, nor for every branch of freshwater study. The items have been described merely so that the student will know what is available for the particular branch he intends to study, and to enable him to devise suitable equipment for his own needs.

A word might perhaps be added here on the actual procedure at the waterside. The absolute novice marches noisily to the scene of operation, splashes his net around for a few minutes, and then complains bitterly that there is nothing in the pond or stream. The more experienced pond-hunter, on the other hand, approaches the water edge cautiously, examining the marginal vegetation carefully for the adult forms of aquatic insects, and for those, such as dragonflies, emerging from their nymphal skin. He treads lightly to avoid causing vibrations which would send all the creatures near the edge of the water out into the middle. Arriving at the water, he scrutinizes the surface carefully for animals which live habitually on the surface film—water springtails, water measurers, water crickets and water skaters. Peering down into the shallow water, he will see beetles, water scorpions and other water-bugs, leeches and perhaps planarian worms crawling over the bottom mud. When he starts wielding his net he takes first a few sweeps along the surface, keeping the open end slightly tilted upwards, and when examining his catch in the net-tube, waits a short time to allow the cloud of rotifers and other plankton to descend into the tube. A sweep or two of the net through the water-plants follows, and only then does he plunge his net down into the water, finally scraping it along the mud surface towards the bank. Only by such a systematic approach will the best results be obtained. To plunge the net here, there and everywhere at the start will merely frighten away the creatures from within reach, and by rendering the water muddy prevent all further observation in the water.

Everyone has his own method of carrying his pond-hunting equipment. Some favour a haversack, some a canvas roll with loops to take the individual items, and others a rigid box, basket, or small attaché case. Personally I prefer the last two, for the various pieces of equipment, together with specimen bottles and tubes, can be packed in them more securely than in a flexible bag. Polythene bags make good containers for the wet nets, and specimen bottles can be slipped into tubes made of corrugated paper to protect them from injury. (Plates 55, L; 58, 6)

On reaching home, the material should be turned out immediately into large flat dishes, even if it is not intended to examine it at once. After a tiring day's collecting, the temptation to leave the material in the tubes or bottles in which they have been carried is great, but it should be resisted, for a delay of an hour or two in transferring the organisms from the narrow tubes will usually prove fatal to many organisms, particularly if they are kept in a heated room.

Most students will, no doubt, wish to study in a living state some of the creatures they have secured, so that a few words on keeping them in aquaria will be of assistance. There is, of course, no need to purchase expensive glass-sided aquaria; much simpler receptacles will serve equally well, if not better. For the large creatures, such as insect larvae, snails and the like, pie-dishes or flat enamel or plastic vessels are very suitable, and a variety of shapes and sizes of these are available. It is advisable to keep each kind of creature to itself, so that a number of dishes will be needed. Most aquatic creatures are adepts at escaping, and a cover, such as a sheet of glass, should be provided for each dish if it has no cover. The individual requirements of each specimen regarding food, light and shelter must, of course, be observed. If possible, the water in the tanks should be pond water, or at least clean rain-water; the chlorination to which much tap water is now subjected is inimical to many creatures.

Animals which undergo a transformation will need to be provided with the means of doing so. Many beetles, for instance, pupate in the mud at the edge of the water, and in captivity these conditions can easily be simulated by heaping a bank of sand or gravel, as the case requires, against the end of the vessel to form a 'shore'. In most cases the larvae will crawl into this bank and pupate successfully. Dragonfly nymphs and other insects which crawl out of the water to transform to the adult stage should, when they are ready to emerge, be kept in a little water at the bottom of deeper vessels and provided with sticks or pieces of reed up which they can climb out of the water.

Smaller creatures, including those of microscopic size, can conveniently be kept in small glass vessels. Clear tumblers, if only partly filled with water, are quite suitable, and petri-dishes, which are very shallow round glass dishes with lids, used extensively in laboratories, are excellent. One of these is illustrated in Plate 64. The important thing to bear in mind in keeping all creatures which need to breathe oxygen dissolved in the water is that the surface area of the water in which they are kept must be great in relation to the depth of the water.

Species that live in streams and rivers need special treatment. They can usually be kept in running tap water if it is not too highly chlorinated, and if particular care is taken to keep the vessels containing them in a cool place. A series of shallow dishes, each covered with gauze or netting, and set in descending levels, provides a good substitute for a stream if a convenient tap and drainage can be arranged.

The food for microscopical creatures need cause no trouble. In most cases a pipette full of water from the pond or stream from which they were taken will provide all they need. 'Cultures' of various microscopic creatures, which can be made by steeping hay or other dried vegetation in water for a day or two, will similarly provide abundant food material.

Use of the Microscope

Sooner or later the student of freshwater life will feel the need for a microscope. There are many good books on the choice and use of a microscope, some of which are mentioned in the bibliography (p. 272), and the reader is referred to these for detailed information. It may be as well, however, if a few general remarks are given here on the subject, particularly on the equipment which is available to aid the examination of microscopical life.

It cannot be stressed too strongly that in purchasing a microscope the advice of a knowledgeable microscopist friend should be obtained if possible. Equally, it must be impressed on the uninitiated that the cheap, new microscopes, often referred to in advertisements euphemistically as 'students' microscopes', are almost useless; and if expense is a consideration, a good second-hand instrument made by one of the reputable firms of microscope manufacturers should be purchased in preference to one of these atrocities. The important part of the microscope is the lower of the two lenses, i.e. nearer the object being studied. This is called the objective and it is usually now identified by its magnification, which is marked on the barrel, e.g. × 10. The upper lens is the ocular (nearest to the eye) and again these lenses are marked with their magnification, e.g. × 5, but it is important to remember that they are only magnifying the image produced by the objective and it is the latter that determines the quality of the image. Most of the older stands take the standard objectives, such as are used on all modern microscopes, and at first a × 2·5 and a × 10 objective will be all that is needed. Later a × 25 or × 40 can be added separately, but an oil immersion × 100 will rarely be required for general freshwater study, and its expense is not justified.

Older objectives which may be picked up second-hand will be marked in focal lengths rather than magnifications e.g. 2 in., $\frac{2}{3}$ in. or 50 mm, 16 mm. The equivalent of these in modern lenses and the magnifications of both when used in combination with different oculars or eyepieces are given in the table opposite.

For freshwater study there is no doubt that a binocular microscope is to be

Approximate magnification with 160-mm tube length

OBJECTIVE			EYEPIECE		
			× 5	× 10	× 15
× 2·5	2 in.	50 mm	12·5	25	37·5
× 10	⅔ in.	16 mm	50	100	150
× 25	¼ in.	6 mm	125	250	375
× 40	⅙ in.	4 mm	200	400	600

preferred. Not only is protracted observation much less tiring when both eyes are used, but the full stereoscopic vision which such an instrument gives, particularly with low magnifications, yields a much more complete idea of the organism, with an appreciation of depth and solidity lacking in a monocular microscope.

One technique which all students of freshwater life should endeavour to perfect is that of 'dark-ground illumination'. This system of lighting, which can most easily be achieved by inserting into a ring below the sub-stage condenser of the microscope a 'wheel-stop' to suit the particular objective in use (Fig. 88), illuminates the specimen by oblique rays only, so that it appears in the microscope

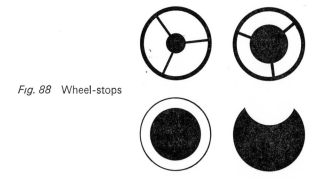

Fig. 88 Wheel-stops

field as a brightly luminous object against a velvety black background. The effect of this mode of illumination on translucent creatures such as bryozoans has to be seen to be believed, and delicate, fine structures such as cilia, which would be difficult to see by ordinary transmitted light, are revealed in all their beauty.

For the examination of very minute organisms under the microscope there are a number of items of equipment which, although not perhaps essential, simplify the work a great deal. One of these is a stage-trough similar to, but smaller than, the pondside trough mentioned on page 239. They can easily be made at home from two pieces of thin glass cemented together with a space between them to give the requisite thickness. A smaller type of trough is that known as 'Botterill's

Trough', illustrated in Plate **60**, C. This consists of two pieces of perspex or laminate, held together by three screws, between which are placed two ordinary microscope slips, separated by a thick rubber band bent double. The advantage of this trough is that it can easily be taken apart for both surfaces of the glasses to be cleaned, and in the event of a glass being broken it can soon be replaced. Stage-troughs are suitable for examining all types of organisms which are attached to weed, such as moss animals, hydra, *Floscularia* and *Vorticella*. For free-swimming creatures other appliances are needed which will restrict their movements. The simplest of these is the live-box (Plate **60, D**), which consists of a flat brass plate with a circular hole in it, over which is mounted a short brass tube. A piece of thin glass is at the bottom, flush with the plate. Another brass tube also holding a thin glass cover-slip fits into the first tube. In use the loose tube is removed, a drop of water containing the organisms is pipetted on to the bottom piece of glass, and the loose tube replaced and slid carefully down until the two pieces of glass trap the creatures, holding them still for examination. Small crustaceans and rotifers can in this way be studied closely.

A refinement of the live-box is the compressorium, one design of which is illustrated in Plate **60**. Here the upper cover-glass is held in an area (swung off-centre in the illustration) which can be lowered by rotating a finely threaded screw. This gives a very fine adjustment, and enables very delicate creatures to be held without damaging them.

Very tiny organisms can be examined if the drop of water containing them is placed on an ordinary microscope slip and a cover-glass lowered on it. To prevent them from being crushed, the *edges* of the cover-glass should, before lowering, be scraped on a piece of Plasticine so that the smallest quantity is left on to raise the cover slightly off the slide. Special cavity slips, with shallow oval or circular depressions in the middle of them, are sold, which can also be used for examining small creatures.

A number of ordinary watch-glasses, obtainable from a jeweller, will be found invaluable to have on the bench when sorting out small organisms for examination. Several pipettes and dip-tubes; a pair of fine forceps for picking up small fragments of weed; and a fine camel-hair brush for handling aquatic worms and small insect larvae, complete the essential equipment for microscopical examination of aquatic organisms. (Plate **60**)

Details of the methods to be adopted for making permanent preparations of aquatic creatures will be found in most books on the use of the microscope.

The equipment used by professional biologists engaged in freshwater research is similar to that which has been described, but there are a few items which are not generally used by amateurs, and a brief description of some of them will form a fitting close to this chapter.

The mention in Chapter 2 of water temperature will have indicated the importance which this factor has on freshwater life, and obviously some means of measuring the temperature at predetermined depths becomes essential in lim-nological research. One method is to use a *reversing thermometer*. This is an accurately calibrated thermometer held in a frame which enables it to swing through 180 degrees, in a vertical direction, when a catch is released. In use the instrument is lowered from a boat to the required depth on the end of a cable. After sufficient interval has elapsed to allow the thermometer to register the temperature, a sharp pull on the cable or the despatch of a weight or 'messenger' down it to release the catch, depending on the particular model, causes the thermometer to reverse, thus breaking the mercury column at the S-bend so that when the instrument is hauled to the surface it still registers the temperature at which it was reversed.

The reversing thermometer has been almost superseded by a thermistor used in conjunction with a galvanometer and a battery (Plate **61**). Variations in temperature cause changes in the electrical resistance of the thermistor. When the 'bomb' is lowered on its cable to the required depth of water, the galvanometer reading enables the temperature to be estimated to an accuracy of plus or minus ·05°. The oxygen concentration of the water is measured with an *oxygen probe*. (Plate **62**)

To obtain samples of water from known depths, a *water bottle* (Plate **61**) is used; this is a hollow cylinder with lids at both ends, and is lowered into the water in the same way as the thermometer. When it goes down the lids are both open, but once it reaches the depth from which a sample of water is required, a 'messenger' is sent down on the cable to release the catch on the lids so that they close tightly and the bottle can then be hauled to the surface with the re-quired sample.

In studying the organisms living on the bottom mud of a lake or large pond, a form of *grab* is employed. This consists of a pair of large toothed jaws, hinged together. It is lowered to the bottom in the open position, the jaws closed by manipulation of the rope and the grab is then hauled to the surface containing the sample of mud for examination. A type frequently used is the Peterson Grab. The dimensions of the grab are such that a known area of the bottom mud is brought up for examination so that, by taking a number of samples from different parts of the lake, it is possible to determine the average numbers of any particular organism per square metre.

Somewhat similar in function to the grab is the piece of equipment known as a *mud-sampler*. This, however, takes not merely a sample of a definite area of the bottom deposit, but a vertical section of the mud to a depth of about 15 cm, and examination of the samples obtained by it enables a great deal of information to be obtained on the chemistry and bacteriology of the surface layers of the mud. (Plate **62**) Deeper samples of the mud are obtained by an apparatus called a *core-sampler*. This will raise an undisturbed core of bottom-deposit several metres in length, and from an examination of the remains of diatoms, pollen, etc. in the various layers of mud, it has been possible to learn much about the age and evolution of lakes.

The importance of aquatic insects in the diet of freshwater fish has led to a good deal of research on the particular insects which are present in certain waters, and their abundance. A convenient way of determining these points has been employed by the research staff of the Freshwater Biological Association, whose laboratories are situated on Windermere. Floating perforated zinc cages of definite dimensions are moored on the surface of the water. The adult insects which emerge from the water, mostly midges of the family Chironomidae, are trapped in the cages, and by periodically removing the cages and counting the winged insects inside, the numbers per square metre can be ascertained. (Plate 59, 4)

During one season between the months of April and November, no fewer than 2,742 specimens of chironomid midges alone, in addition to many dragon-flies, mayflies and caddis-flies, were collected from two of the cages illustrated, which were floating on the surface of a small Westmorland tarn. The midges belonged to fifty species, of which two were new to science and two were only previously known from single specimens. This is striking evidence of the vast numbers of such insects which are present in fresh water, and also suggests a simple way in which the amateur could carry out similar investigation on ponds near his own home.

19: FRESHWATER BIOLOGY IN THE SERVICE OF MANKIND

In preceding chapters, as the various groups of freshwater organisms have been discussed, brief mention has been made from time to time of the impact of some of them on human life. It may be of interest if, in this final chapter, that aspect of the subject is enlarged upon, and a few instances given of the way in which freshwater biology is serving mankind.

Medical Applications

The medical aspect comes first to mind, and perhaps the most important contribution which freshwater biology has made in this field is the elucidation of details of the life-histories of creatures, not all of them insects, which either act as intermediate hosts of parasites harmful to man or his domestic animals, or are themselves parasitic. This knowledge has permitted a control to be applied to the worst pests at the most convenient, or most effective, stages of their development, and in this way has rendered many parts of the world, which were formerly almost uninhabitable, thriving centres of population, agriculture and commerce.

Thus in those areas where a determined effort has been made malaria, the most serious of all the diseases of the world, judged from the standpoint of number of victims, has been largely eradicated by the adoption of measures resulting from a close study of the biology of the anopheline mosquitoes that serve as the intermediate hosts of the protozoan parasites *Plasmodium*. The draining of waters in which the larvae are found, or the covering of the surface of the water with a film of oil to kill them off, are the classical methods which have been in use for many years.

More recently another approach to malarial control has been made by careful study of the species *Anopheles maculipennis*, which transmits malaria in Europe. It was known that, although this species was widespread, malaria did not necessarily occur in the areas where it was found. An amateur entomologist in Italy, who bred anopheline mosquitoes as a hobby, noticed that the eggs of *A. maculipennis*, which he collected from different localities, varied in colour and pattern but were of five general types. The larvae and adults bred from the eggs were apparently indistinguishable from one another, yet the same females always laid the same type of egg. Investigations extended over other parts of Europe showed that the same situation was general, and it became clear that five distinct

R

races of *A. maculipennis* exist, only three of which will, in a state of nature, transmit malaria to human beings. In Europe, therefore, the disease only coincides with the distribution of one or other of these races. Thus the observations of the amateur biologist have helped materially to solve a problem which had puzzled malaria control officers for some time and have provided a sound basis for the control of the disease.

Schistosomiasis, a dreaded and painful disease, which in some parts of the world has been said to affect every third person, is caused by a small parasitic flatworm, *Bilharzia* or *Schistosoma*, which passes its larval stages in freshwater snails and in a later stage enters the human body through the skin when the victim is washing, bathing or drinking the water. It was observed that in its free-swimming stage the larval fluke died if the water in which it was living was drawn and kept quite still for thirty-six hours. Thus a ready means was provided, for those who would trouble to adopt it, of preventing infection.

These are only two examples of how quite simple observations on freshwater organisms have benefited humanity.

Fortunately, in this country we are spared most of such scourges, although malaria, under the name of 'ague', was a common disease in some coastal districts in former days. Much of the research on tropical medicine which has benefited other countries has, however, been carried out by British-trained biologists.

Food Production

This country has neglected the great contribution which the fresh waters can provide in the way of food. In most other parts of the world freshwater fish are consumed in large quantities and reared artificially on a considerable scale. The many problems involved in this form of pisciculture—increase in the rate of growth, control of disease and so on—have been pursued diligently by freshwater biologists, many of whom were trained in this country, but little of the knowledge has been applied to commercial fish production here.

The abundance of sea-fish around the coasts of Britain, and the ease with which they can be made available to any town or village in the country, have caused us to neglect the cultivation of our freshwater fish, although there is little doubt that this could be carried out just as successfully here as on the Continent, providing the prejudice against coarse fish could be overcome. The individual angler, of course, eats and enjoys the fish he catches, but it would be a different matter to persuade the general public that coarse fish are palatable and so make commercial exploitation practicable. Only the salmon, and, to a lesser extent, trout and eels, are readily purchased by urban populations.

The fish most suitable for culture in ponds is carp, but on the Continent tench and pike are also introduced into fishponds, the former to exploit the pond bottom and the pike as a 'policeman' to keep down the numbers of small fish, such as roach and perch, which would otherwise compete with the carp for the available food.

In fertile ponds the fish are not given any additional food, but in most waters artificial feeding is usually necessary, and this takes two forms: food directly intended for the fish; and material to effect the increase of the pond plankton organisms, on which the fish feed, either indirectly or directly. Various substances are used as fertilizers for this purpose, including phosphorus compounds, bone meal and plain horse manure.

Good carp ponds will produce annually up to 280 kg of marketable fish per hectare, and since very little labour is required in their maintenance, they provide a valuable addition to the table, even if maintained only on a small scale. By introducing carp and *Tilapia* into ponds, the full potential of the habitat is realized, the carp feeding on bottom invertebrates and *Tilapia* on plankton, with a great increase in productivity.

The U.S.A., Germany and Poland are countries in which a good deal of attention has been given to fish culture. In the last country, for instance, the inland fisheries, which are state-controlled, are based on 300,000 hectares of lakes, 80,000 hectares of artificial carp ponds and 25,000 hectares of rivers. (A hectare is equivalent to 2·471143 acres.) The natural output of fish is 30 kilogrammes per hectare in lakes, 50 kg/hectare in rivers and 100 kg/hectare in carp ponds. Most carp ponds, however, are improved by mechanical means (ploughing, removal of excess plants and fertilizing) and by artificial feeding of the fish, mostly with rye, maize and legumes. The average crop of fish may then be as much as 1,500 kg/hectare or about four-fifths of a ton per acre per annum in favoured ponds.

The most important fish cultivated are bream, tench, pike and roach. Perch is numerous but is regarded somewhat as a pest. Coarse fisheries are estimated to yield about 43,000 metric tonnes per annum, lakes providing 9,000 tonnes, carp ponds 32,500 tonnes and rivers 1,500 tonnes. Angling for coarse fish is very popular in Poland, both in rivers and lakes. Fishing licences are, however, only obtainable after appearing before a local committee of professional biologists and fishery experts and passing a verbal examination on fish and the law relating to fishing!

The figures above will perhaps indicate what a valuable addition to the national diet could be brought about by the exploitation of our freshwater resources—an exploitation which would do nothing to spoil their beauty or use for other purposes, but rather the reverse, for pollution would not be tolerated.

Algae Culture

Fish are not the only items of commercial importance which fresh water is capable of giving, and investigations carried out under the stress of war have pointed the way to another method of exploitation. In the process of photosynthesis, most plants produce starch as a reserve material, but in some, notably certain diatoms and green algae such as *Chlamydomonas*, oil or fat globules are formed. The suggestion has, therefore, been made that large quantities of such organisms should be cultivated in simple nutrient solutions and under optimum light conditions. The reserve material which they would produce could then be extracted to provide fats and oils which the world at present needs so badly.

Preliminary experiments which were carried out in Germany during the 1939–45 war showed that the suggestion was not so far-fetched as it might at first seem. The algae were cultured in nutrient solutions contained in glass cylinders about 1 metre high and 5 cm in diameter. They grew luxuriantly if the solution was well aerated, and in about a fortnight underwent a 'fatty degeneration', when the cells became packed with large globules of oil. These experiments showed that banks of such tubes should yield at least twice as much fat per unit area of culture medium as that obtained with the usual oil-seed plants. No doubt with other strains of algae and the use of artificial illumination to speed up the process, the yield of fat could be increased considerably. There is little doubt that in this way large quantities of fat or oil could be produced, but whether as a commercial proposition has yet to be determined. Nevertheless, the investigation indicates a further possible application of freshwater biology to human welfare.

Water-supply

Man has progressed a great deal since his early ancestors sited their dwellings near a lake or stream, but he still has need of the precious substance which influenced his choice of habitat then, and one of the first essentials of a civilized community is that an abundant and pure water-supply be available. The vast quantities now consumed by urban populations may be gauged by the fact that London alone requires some 400 million gallons daily on the average throughout the year. The provisions of such staggering quantities of the purity now demanded is a responsibility of the biologist no less than the engineer.

We are all prone to take too much for granted our good fortune in having an abundant and pure water-supply, and we waste water in a manner which cannot but appal anyone who is aware of the effort and care which must be given to provide us with it. Scarcely a week passes without our hearing of some scheme for flooding this valley, of damming that river or of building yet another artificial reservoir, and we may begin to wonder whether this insatiable demand to provide more and more water would be necessary if we devoted more thought to the design of taps that did not leak, to the storage of the rain-water from our own houses, and most of all to the education of the public so that the wastage of water might be considered a serious offence. A knowledge of the problems of the water engineer would go far towards this end.

The essentials of a good water-supply are that it should be abundant, of good physical quality, free from harmful substances, without objectionable taste and smell, and—perhaps most important of all—containing no bacteria which might endanger public health.

We need say little about the last aspect. In Britain, at least, outbreaks of typhoid and other diseases caused by waterborne bacteria are sufficiently rare to be front-page news when they do occur. But it was not always so. In earlier days, epidemics which we now know to have been caused by water contamination were taken as a matter of course, and why they were frequent can well be understood when we remember, for example, that as late as the nineteenth century one of the intake pipes from which part of London's supply was taken from the Thames was situated within a few yards of the outlet of the main sewer. The recognition of bacteria as the causative organisms, the constant vigilance of bacteriologists and the chlorination of most supplies have almost eradicated this problem here. It is the provision of water that is clear and free from objectionable smells and tastes which gives the waterworks biologists their biggest problem.

Those undertakings which are fortunate enough to draw their water-supply either from deep boreholes and wells, or from natural lakes set on hard rocks— the oligotrophic types of lake mentioned in Chapter 2—have relatively few biological problems, other than control of the reservoir areas.

In many parts of the country, however, particularly in the southern and eastern parts of Britain, the water is drawn either from rivers or from soils rich in nutrient salts. Such waters, when stored in reservoirs, produce an abundant plankton, both flora and fauna, especially at certain times of the year; and if no control were exercised, they would quickly develop objectionable smells and tastes, and the large number of organisms in them would quickly block the purifying filters. The constant vigilance of the biological staff is needed to prevent such occurrences. Large reservoirs, such as the Queen Elizabeth II reservoir of the Metropolitan Water Board at Molesey, and the Wraysbury reservoir, although man-made, are sufficiently extensive and deep to exhibit all the characteristics of a typical eutrophic lake and, as was mentioned in Chapter 2, the results of the investigations which have taken place in recent years on the life-cycle of such lakes are being applied daily by water undertakings.

The numbers of microscopic plants and animals in a reservoir, like those in a lake, vary throughout the year, and the particular kinds abundant at any one time also change. Towards the end of March or early April diatoms in the water, such as *Asterionella*, *Stephanodiscus* and *Fragilaria*, multiply rapidly and appear in vast numbers. Some idea of their abundance may be gauged from the fact that on one occasion it was estimated that the total dry weight of *Fragilaria* in one London reservoir with a volume of 750 million gallons, drawing water from the Thames, was 110 tonnes, and about a tonne of the diatoms would need to be removed by the filters daily from the water.

The diatom outburst is usually followed by a great growth of green algae such as *Eudorina*, desmids and filamentous forms such as *Cladophora*, or 'blanket-weed', and about this time there may be a great increase in the number of microscopic animals—crustacea and the like. In autumn the blue-green algae are at their maximum, and a second increase of diatoms and green algae may also occur then.

The mechanical problems involved in filtering out these vast quantities of algae are great. Diatoms, in particular, by reason of their hard, indestructible silica frustules, quickly choke the filters, and the attentions of the waterworks biologists are directed to means, firstly of limiting the growth of algae, and secondly, of preventing large quantities reaching the filters. Throughout the year, water samples are taken and analysed chemically to find out the quantities of nutrient salts present in the reservoir. From this information and from past experience, it is possible to forecast what organisms to expect and to gain some idea of the probable abundances. Remedial measures, such as the use of an algicide or algae poison, can then be applied before the maximum outburst occurs.

Copper sulphate is the algicide most generally used, but potassium permanganate and activated carbon are also employed on occasions, as they have certain other valuable properties apart from killing off the algae. The old method of applying the algicide was to place the solid chemical in coarse bags which were towed back and forth across the reservoir from the stern of a boat. The actual concentration used is of some importance, for some organisms need more than others to kill them off, and care must be taken not to kill other creatures in the water. With copper sulphate it is usually between 0·25 and 0·5 part per million (2·1 to 4 lb. of the chemical to every million gallons of water). The degree of concentration could be controlled by varying the speed at which the bag was towed through the water. The 'bag-dragging' method has been largely replaced by the use of modern power-spraying equipment.

Substances known as coagulants are sometimes used instead of algicides. These materials, an example of which is aluminium sulphate, form into a gelatinous flocculent mass on coming into contact with water. In this the algae become entangled and sink to the bottom.

Organisms are not uniformly scattered throughout the water, but are usually congregated at particular levels in the water, depending on their kind. It is sometimes possible to draw off water from a layer which is relatively free from algae, and for this reason most reservoirs, when constructed, have provision for drawing off the water at different levels. Examination of samples of the water taken from different depths enables the water engineer to decide from which level he shall draw.

From the reservoirs, the water passes to the filters, and in most waterworks there are two kinds, the primary and secondary filters. (Plate 63) In the first the beds are of coarse sand and the water passes through fairly quickly—leaving only the larger organisms and particles behind. Rotary strainers are sometimes used in addition. The secondary filters, which are much larger in area, contain finer sand, and through them the rate of filtration is much slower. The efficiency of the filters is due not merely to the physical action of the sand but to the growth of a biological film, consisting mainly of bacteria and algae, which grows on an established filter bed. This living film—the so-called *zooglea* layer—coats the sand grains and forms a filter capable, not only of straining out almost every living particle, but also of purifying the water as it passes through by oxidation of ammonia. The dissolved organic matter in the water is converted by the bacteria into simple inorganic salts, and the green algae, by their photosynthesis, convert the carbon dioxide into oxygen to help the process of purification.

In time, however, the film becomes thick and would eventually stop the flow of water altogether. The primary filters are then cleaned by drawing off the water and forcing a stream of compressed air through the filter layer from below, the debris then being washed away by a stream of water into side channels. The secondary filters, after being emptied, have the top layer of sand carefully skimmed off and a new film is allowed to grow before water from the filter is allowed into the consumers' supply.

Although chlorination of the water sometimes takes place before filtration, only small concentrations, just sufficient to kill off the algae, are used at this stage. If larger quantities were introduced into the water, they would be lethal to many of the organisms in the zooglea layer of the filters and destroy their efficacy. Thus the heavier chlorination necessary to destroy any bacteria that may be left after slow sand filtration, or that may gain access later, is applied after filtering and the now pure, sparkling water passes into the mains for delivery to consumers.

There are a number of other creatures of interest to the freshwater biologist which are found in waterworks but which rarely trouble the consumer. Freshwater sponges, moss animals and molluscs, particularly the zebra mussel, *Dreissena*, may grow abundantly in the pipes and by restricting the flow cause considerable trouble to the engineers.

Sewage Disposal

It has been estimated that in Britain alone over 3,000 million gallons of water are distributed daily by water authorities. In addition, large supplies are taken by industrial concerns direct from rivers and other sources. Most of this vast quantity of water finds its way eventually into surface waters as sewage effluent and were it all untreated it would give rise to many problems of river pollution and hygiene.

It is unfortunately true that the state of many of our rivers and shores indicates that much remains to be done in the treatment of sewage and industrial effluents, but a good deal has been achieved and most inland towns of any size now adopt one or other of the biological methods of purification which result in the material finally discharged into river or sea being almost pure water, containing only inorganic salts. It is perhaps not generally realized that the process of sewage purification, after the insoluble solids have been removed, depends mainly on biological principles, and that a knowledge of freshwater biology is needed by those whose duty it is to carry out this work. The essence of the method is the purification of the sewage effluent by oxygenation.

In older sewage works the crude sewage is taken first to deep but narrow pits called grit-pits, where gravel, stones and similar materials which have been collected by the sewage streams settle. It then passes through screens, which hold back paper and other fibrous matter and which also serve to break down the other solids into a finely divided state. The now turbid liquid is next allowed to settle for some hours, often with some limy substance added to it, in sedimentation tanks, where much of the fine solid matter settles and may be subsequently removed for use as manure. From there the now much clearer liquid is taken to the percolating filters over which it is sprayed by means of sprinklers travelling over the surface. So far the purification has been purely physical in character, but the percolating filter does its work by biological means.

The percolating filter, or bacteria bed as it is sometimes called, is filled with stones, coke or clinkers to a depth of about 2 metres. When first made it is ineffective, but after some time, possibly only after a month or two, the surface of the stones becomes covered with a zooglea layer (somewhat different in character from that of the filters in a waterworks), and is then ready for use.

Apart from a certain amount of non-living matter, the layer is made up of a complex association of bacteria, algae, fungi and Protozoa, to the action of all of which, in converting undesirable substances into simple, harmless materials, the filter owes its efficiency. The Protozoa are of particular interest. They are mostly ciliates (see page 72) of both the fixed and freely moving forms. It will be remembered from the description of these animals that they create currents in the water by means of vibrating cilia, and both move about and obtain their food by this means. They can cause small particles in the water to clump together by means of a secretion which they put out, and in the sewage filters they act not only by ingesting particles from the water, but also by causing bacteria to clump together and fall out of suspension, thus rendering clearer the effluent passing through the filter.

As time goes on the zooglea layer becomes thick with the growth of organisms on it, and if no other causes were operating, it would soon become not a filter but a seal, effectively preventing the passage of the sewage liquid. Fortunately, however, another biological phase now comes into operation to prevent this from happening. The very profusion of the mat produced by the algae bacteria and fungi attracts larger creatures, which habitually feed on such detritus, the most important of which are oligochaete worms and the larvae of several families of flies belonging to the order *Diptera*. These live both in the filter and on the surface, feeding on the film and breaking it up, so that small pieces get washed through. In this way the film is prevented from forming a close mat. In spring, when the film would be especially subject to this state of affairs owing to the prolific growth of the plants and the accumulation of much dead material, the animal life is particularly active, so that the layer is broken up to a marked extent, a phenomenon called by the sewage engineer the *vernal slough*.

The worms belong mostly to the family Enchytraeidae, and *Lumbricillus lineatus* is perhaps the commonest species found, often forming dense red masses made up of many individuals. *Tubifex* is also found at times in the filters.

Of the insect larvae, the kinds most abundantly found are the moth or sewage-flies, *Psychoda* (p. 201), and some idea of the vast numbers of larvae which live in the filters may be realized when it is said that in summer it has been estimated that some 40,000,000 adults emerge per day off one acre (0·4 hectares) of filter. They appear as a dense cloud over the whole filter area and, although not long-distance fliers, are carried by the wind or air currents to houses in the vicinity of the sewage works, much to the discomfort of the people living there. So far as is known, the insects do not spread disease, neither are they attracted by human food as are house-flies, but, nevertheless, the vast numbers in which they appear make them a considerable nuisance. A good deal of research has been directed to the control of the flies without adopting such drastic means that the services of their larvae are lost or the other creatures in the zooglea layer killed. The use of modern insecticides has proved effective to some extent.

A more modern method of sewage purification which is now used extensively is that known as the *activated sludge* system. In this, the sewage is allowed to flow down narrow channels where it is violently agitated, either by moving paddles or by having compressed air forced into it. This agitation makes the liquid frothy and the solids in suspension become flocculent masses, which on examination are found to be largely bacteria and Protozoa, the latter including *Vorticella* and *Campanella*. The activated sludge, consisting of these flocs, passes to settlement tanks, where the masses separate out and the clarified effluent passes on. Some of the activated sludge is then passed back to the incoming sewage stream to ensure that there are sufficient purifying organisms to continue the process. The flocs in the activated sludge system perform the same function of purification as the zooglea layer on the percolating filter.

From both processes the final effluent is a clean liquid with the solids removed, the harmful substances oxidized, and in this stage it can be allowed to pass into a stream or the sea without danger of pollution.

A by-product of sewage purification is the release of the gas methane or 'marsh-gas', which, as was mentioned on page 49, is produced by the action of bacteria in conditions where there is much decomposition going on and where oxygen is deficient. The purification plant at Mogden, in Middlesex, produces enough sludge gas, containing seventy per cent methane, to supply a town of 100,000 people with light and heat. The gas is, however, used on the plant for providing power for heating the main buildings and for operating the works vehicles. The amount of fuel-oil saved by the use of the gas is estimated at over 1,250,000 gallons per annum which, considered in terms of cash, is no mean contribution from organisms that are usually regarded as troublesome!

During the 1939–45 war, compressed gas from this plant was used to fill several million heavy incendiary bombs (the methane being dissolved in petrol), which were used with devastating results on Germany. As a consequence, Mogden itself became a target for German bombers.

Pollution

The pollution of natural waters by man is not a new phenomenon. Ever since he inhabited the margins of streams and lakes, the water has been the repository for his waste products, but it is only within the last century that increasing population and large-scale industrialization have made it a serious problem nationally, and attracted attention to its biological effects.

The condition of many of our rivers has been investigated by a succession of Royal Commissions and the state of some of them during the last century is strikingly illustrated in the published reports. The River Pollution Prevention Commission, 1868–74, for example, recorded that a perfectly legible letter received in the course of obtaining evidence was written with water taken from one river!

Some kinds of pollution, however, are new and these include powerful, persistent poisons used in pesticides such as the organochlorine compounds that can cause disastrous effects on aquatic plants and animals over a wide area. Toxic substances in solution also reach rivers as effluents from many industries and include salts of heavy metals such as zinc and lead, as well as acids, alkalis and phenols. All these can kill off the living organisms for long stretches of river below the point of entry.

From some industries, such as mining or quarrying, suspensions of fine particles reach rivers. If the particles are light enough to float they make the water turbid and prevent light from reaching plants. Not only do these die, but so do the animals dependent on them for food or shelter. If the particles are heavy, and settle on the bottom, they fill up the spaces between stones and deprive creatures that live under stones of their habitat. Mussels may be smothered, but burrowing or tube-building animals such as chironomid larvae and aquatic worms are usually not affected.

Organic matter in suspension is perhaps the most universal pollutant and the commonest source is untreated domestic sewage, although paper and textile mills provide their quota in some areas. The immediate effect of these substances is to use up the dissolved oxygen in the water for some distance downstream. As bacteria break down the organic compounds, using up much oxygen in their respiration, they increase considerably for a time and then decline as the substances are converted. 'Sewage-fungus'—a term used for aggregations not only of various kinds of fungi, but also colonial bacteria, including the evil-smelling sulphur bacteria, and Protozoa such as the ciliates *Carchesium* and *Epistylis*—become visible as white, pinkish or brownish masses on solid objects in the river. A later stage is the abundant growth of filamentous algae, especially blanket-weed, *Cladophora glomerata*. As the nutrients on which these organisms feed are used up, the fungi, algae and Protozoa decline in succession. Most invertebrate animals and fish disappear or become reduced through lack of oxygen and the blanketing effect of the mats of sewage fungus and algae, but tubificid worms, chironomids and the water-louse, *Asellus*, survive.

A term much used today is eutrophication. The natural eutrophication of lakes was described in Chapter 2, but the current use refers to the rapid enriching of both lakes and rivers by nutrients, including phosphates and nitrates, derived from organic pollution, and, in some areas, from seepages from agricultural land that has been dressed with artificial fertilizers. The situation in Britain has not yet reached so serious a state as it has in other countries. The turbid and un-healthy state of the Great Lakes in North America has caused much concern in recent years, as have some lakes in Switzerland. In Lake Zürich, for example, which receives nutrient salts from sewage and industry, the colour of the water has changed from blue to yellowish or brownish-green through the great in-crease in plankton. The oxygen content of the hypolimnion (p. 22) has been reduced so seriously that water supplies for the surrounding area, which is drawn from the deeper parts of the lake, are badly affected.

As lakes become more eutrophic the species of plants and animals undergo an ecological succession. Diatoms are succeeded by blue-green algae; rooted plants increase and shores may become covered with a dense mat of filamentous algae, mainly *Cladophora*; salmonid fish give way to perch and later bream, roach and rudd. The algal blooms from blue-greens may be poisonous to fish and even to cattle drinking the water. They also cause bad smells and tastes, reducing the value of the water for public supplies and making it unpleasant for amenity purposes. Once started, the process of eutrophication seems irreversible in lakes.

Although the scientific study of polluted waters has been carried out in Britain, since about 1890, by river boards and others, the emphasis was originally on chemical and bacteriological aspects as being of more urgency. The awakening of interest in freshwater biology in the 1930s stimulated research into wider biological problems, and in the 1960s increasing concern for man's environment generally brought the subject of pollution to public notice. Increased pressure from anglers, especially through the activities of the Anglers' Cooperative Association in bringing legal action in the courts against large organizations responsible for pollution of rivers, and other public expressions of concern, have resulted in some progress being made towards cleaner rivers, and fish have returned to some for the first time for many years. It is to be hoped that the regional water authorities set up in 1973 will tackle the problem of pollution effectively and restore our rivers to an acceptable state.

Freshwater Research

Freshwater problems of one kind or another thus occur in such a wide variety of human undertakings that investigations in freshwater biology now occupy an important place in the research programmes of all countries.

In Britain several research laboratories are concerned with the subject, but from our point of view, perhaps the most interesting and important of these is that maintained by the Freshwater Biological Association. This Association was formed in 1929 as a result of discussions which had taken place at the meeting of the British Association for the Advancement of Science, held at Glasgow in 1928. The declared objects of the Association were '*the promotion and investigation of the biology (in the widest interpretation of the word) of the animals and plants found in fresh water*'. The first research station was opened in 1931 at Wray Castle on the shores of Windermere, and since that time a considerable amount of original research on many aspects of freshwater biology has been carried out by the team of specialist scientific workers on the staff. To provide new premises, the old Ferry Hotel on Windermere was purchased in 1948, and the laboratory transferred there in 1950. It is at present supported by the Natural Environment Research Council, but its income is augmented by subscriptions from the principal water undertakings, universities, fishery boards, angling clubs and other organizations, as well as private individuals. Membership is open to all interested in or concerned with freshwater biology.

The staff includes research workers specializing in many fields of zoology, botany, chemistry, physiology, bacteriology and hydrology. Naturally the problems investigated cover a very wide field. Much long-term research has, for instance, been carried out on the physical and chemical conditions obtaining in lakes—the formation of the thermocline (p. 22), the effect of the wind on the water surface; the variations in the supply of chemical substances in lakes and the chemical reactions which occur in the mud. One of the most important subjects on which work has been carried out continuously since the laboratory was established is the investigation of factors which control the growth of algae under natural conditions. This has necessitated, not only much detailed chemical analysis of water samples, but also culture in the laboratory of algae, including diatoms, under controlled conditions as regards nutrients and light.

The vital part played by bacteria in the water has been indicated from time to time in the preceding chapters, and this aspect has been studied closely by the bacteriological staff of the Association.

Since the larvae of many aquatic insects, such as stoneflies, mayflies and caddis-flies, form such a large part of the food of some freshwater fish, it is important for members of fishery boards and others to be able to identify them accurately. Although the adults of these insects can be readily identified, little had previously been done to determine with certainty the identity of all the larvae or nymphs. At the Association's laboratories large numbers of the immature stages of these insects have been reared to maturity and, as a result of this work which has been carried out over a number of years, much knowledge of the larval and nymphal stages has been gained, and it is now possible to identify many of them. Identification keys for some of the groups, published by the Association, are referred to in the bibliography on pages 270 to 272.

Finally, much research has been conducted by the Association on fish—their food, rate of growth, spawning and other matters—to add to the knowledge of the biology of the main freshwater species.

Only a few of the lines on which research is proceeding have been mentioned, but summaries of investigations carried out during the previous year are contained in each of the Association's Annual Reports, which thus serve as very valuable résumés of freshwater research in this country.

In 1964 the Association opened a River Laboratory at East Stoke, in Dorset, to carry out comparable research, but with special application to running waters.

Undoubtedly, many of the basic problems of freshwater biology can be tackled only by teams of specialists working in such laboratories as those of the Freshwater Biological Association. It would be a mistake, however, to assume that the subject is outside the scope of the amateur and the lone worker. Much remains to be discovered about many aquatic animals and plants. Anyone who can make himself familiar with a group of organisms, mastering first the correct identification of the species and then proceeding to their biology, and especially the ecological conditions under which they live, may well produce results of scientific value.

★   ★   ★

Fresh water always has been and will remain one of man's most important natural resources—too long taken for granted. Increasing human populations make it essential that our freshwater habitats are conserved wisely and unpolluted for the good of all. They must be regarded not only as sources of supply for domestic and industrial purposes nor yet as convenient places in which to dispose of waste products. Their value as recreational and educational amenities must be realized. In this connection, the continuing disappearance of many small ponds is of particular concern (p. 28). The attempts of some county and regional nature conservation trusts to save ponds of special educational value, or ecological interest, is encouraging and it is to be hoped that their example will be followed throughout the country. All of us who find interest and enjoyment in studying the living things in fresh water have an especial responsibility in widening this interest among those around us and so building up an informed public opinion that will do much to ensure that these places of delight that *we* have enjoyed are preserved unspoilt for the use of posterity.

The words of Gerard Manley Hopkins express the feelings and hopes of all those who have loved the wild, wet places where are found the living things that have been described in this book:

> What would the world be, once bereft
> Of wet and wildness ? Let them be left,
> O let them be left, wildness and wet;
> Long live the weeds and the wildness yet.

SUMMARY OF THE CLASSIFICATION OF MAIN GROUPS OF AQUATIC ANIMALS

Phylum: **PROTOZOA** Non-cellular animals
    Class *MASTIGOPHORA* (Flagellata) *Euglena, Ceratium,* etc.
    Class *RHIZOPODA* (Sarcodina) *Amoeba, Arcella*
    Class *ACTINOPODA Actinophrys*
    Class *CILIOPHORA Paramecium, Vorticella*
    Class *SPOROZOA* Parasitic Protozoa

Phylum: **PORIFERA** Sponges
    Class *DEMOSPONGIAE Euspongilla*

Phylum: **COELENTERATA** (Cnidaria)
    Class *HYDROZOA* Hydra

Phylum: **PLATYHELMINTHES** Flatworms
    Class *TURBELLARIA*
        Order RHABDOCOELA *Dalyellia*
        Order TRICLADIDA *Planaria*
    Class *TREMATODA* Flukes
    Class *CESTODA* Tapeworms

Phylum: **ASCHELMINTHES** Worm-like animals
    Class *NEMATODA* Roundworms
    Class *NEMATOMORPHA* Hairworms
    Class *ROTIFERA* Rotifers
    Class *GASTROTRICHA* 'Hairy backs'

Phylum: **ACANTHOCEPHALA** Proboscis roundworms

Phylum: **ECTOPROCTA** Moss animals
    Class *PHYLACTOLAEMATA Cristatella, Plumatella*

Phylum: **MOLLUSCA** Molluscs
    Class *GASTROPODA* Water snails
    Class *BIVALVIA* (Lamellibranchia) Mussels and cockles

Phylum: **ANNELIDA** Segmented worms
    Class *OLIGOCHAETA Stylaria*
    Class *HIRUDINEA* Leeches *Haemopsis*

Phylum: **ARTHROPODA** Jointed limbed animals
    Class *CRUSTACEA*

Sub-class *BRANCHIOPODA*
    Order ANOSTRACA 'Fairy shrimps', *Chirocephalus*
    Order NOTOSTRACA Apus, *Triops*
    Order CLADOCERA *Daphnia*
Sub-class *OSTRACODA Cypria*
Sub-class *COPEPODA Cyclops*
Sub-class *BRANCHIURA* Fish-lice, *Argulus*
Sub-class *MALACOSTRACA*
    Order PERACARIDA
    Order MYSIDACEA *Mysis*
    Order ISOPODA Freshwater louse, *Asellus*
    Order AMPHIPODA Freshwater shrimps, *Gammarus*
    Order EUCARIDA
    Sub-order DECAPODA Crayfish, *Potamobius*
Class *INSECTA* Insects
Sub-class *APTERYGOTA*
    Order COLLEMBOLA Water springtails
Sub-class *PTYERYGOTA:*

        EXOPTERYGOTA
    Order EPHEMEROPTERA Mayflies
    Order ODONATA Dragonflies
    Order PLECOPTERA Stoneflies
    Order HEMIPTERA Water Bugs

        ENDOPTERYGOTA
    Order NEUROPTERA
    Sub-order MEGALOPTERA Alder-flies
    Sub-order PLANIPENNIA Lacewings, sponge-flies
    Order COLEOPTERA Beetles
    Sub-order ADEPHAGA *Gyrinus, Dytiscus*
    Sub-order POLYPHAGA *Hydrophilus*
    Order TRICHOPTERA Caddis-flies
    Order LEPIDOPTERA China mark moths
    Order DIPTERA Two-winged flies
    Order HYMENOPTERA Ichneumon flies, *Agriotypus*
Class *ARACHNIDA*
    Order ARANEAE Water spider
    Order ACARI Water mites
Class *TARDIGRADA* 'Water bears'

Phylum: **CHORDATA** Vertebrates
Class *MARSIPOBRANCHII*
Sub-class *CYCLOSTOMATA* Lampreys
Class *PISCES* Bony fishes
Class *AMPHIBIA* Newts, frogs and toads

HYDROGEN-ION CONCENTRATION AND pH

References to the pH and hydrogen-ion concentration of the water have been made from time to time throughout this book. The following brief and simplified account of the meaning of these terms, and their application to fresh water, may be of interest to the serious student.

According to the theory of electrolytic dissociation, water, and all liquids containing water, contain free, positively charged hydrogen (H +) ions and negatively charged hydroxyl (OH −) ions. If there is an excess of hydrogen ions, the liquid will be acid, if there is an excess of hydroxyl ions, it will be alkaline, and if the numbers are equal, the liquid is said to be neutral. It has been found by methods of electrical conductivity that at a temperature of $18°$ C ($= 64.4°$ F), a litre of pure water contains one ten-millionth of a gramme $\left(= \text{0.0000001, 1} \times \text{10}^{-7} \text{ or } \frac{1}{10^7} \right)$ of ionized hydrogen and an equivalent amount of ionized hydroxyl. The product of the concentration of these two ions (i.e. at $18°$ C 10^{-14}) always has the same value at the same temperature, so that if the concentration of only one of the ions is determined, that of the other is also known. Thus, if the concentration of the hydrogen ion is 10^{-5}, that of the hydroxyl will be 10^{-9}. In practice, only the concentration of the hydrogen ions is taken into consideration, and since the figures 10^{-5} and so on are somewhat clumsy, as is also the term 'hydrogen-ion concentration', the symbol pH is employed, which merely indicates the logarithm of the reciprocal of the hydrogen-ion concentration. Since for a neutral solution the hydrogen-ion concentration is 10^{-7}, or as it may be written $\frac{1}{10^7}$, the pH for such a concentration is thus simply given as the index of the power, that is, 7. In an acid solution in which the hydrogen-ion concentration increases to, say, 10^{-6}, the pH value decreases to 6, and in an alkaline solution where the *hydroxyl* ions increase to 10^{-6}, there will only be 10^{-8} hydrogen ions, so that the pH value will be 8. The pH value, therefore, decreases below 7 with acidity and increases above 7 with alkalinity, and it should be remembered that this is the converse of the way in which the hydrogen-ion concentration really varies.

The pH of any natural water depends almost entirely on variations in the amount of carbon dioxide and carbonates, including bicarbonates, it contains. The carbon dioxide tends to make it acid and the carbonates tend to make it alkaline, but owing to the 'buffer' effect of the carbonates, comparatively large variations in carbon dioxide make only small changes in the pH when such carbonates are present, and in 'hard' waters, that is those rich in carbonates, the variations in pH are less throughout the year than in soft waters. The principal carbonates in fresh water are those of calcium, which are normally present as bicarbonates, and where the alkalinity is known to be due to the presence of calcium, the pH reaction will run parallel with the quantities of calcium in the water. By making a pH determination, a good idea is thus obtained of the amount of calcium available, which is very useful information to have when studying the ecology of creatures which are greatly influenced by the presence or absence of calcium, such as snails, crustacea, etc.

During the hours of daylight, when plants are actively photosynthesizing, carbon dioxide is removed from the water, making it less acid and increasing the pH, and in a very weedy pond or ditch on a sunny summer day the pH of the water may rise as high as 9 or even more.

Conversely, when large quantities of carbon dioxide and organic acids are produced, as, for instance, by decomposition on the bottom of a pond, the water tends towards acidity and has a low pH. Acid peaty moorland pools will often have a pH of 5·6 or lower.

In general, therefore, in average waters, the pH will be highest in summer when photosynthesis is rapid and lowest in winter when much of the vegetation of the past summer is decomposing, but it will vary from day to day, from hour to hour and from place to place. Typical values which were taken in one locality were: for weedy ponds in summer pH 8; stagnant ditches rich in decomposing matter 7·2; ponds containing much animal life and only an average number of plants 7·8.

The importance of pH values lies in the fact that animal tissues in contact with the water are sensitive to changes in the hydrogen-ion concentration. This is particularly so in the case of microscopical creatures such as the Protozoa and also flatworms. *Paramecium*, for instance, will die when the pH is as high as 8·4 and *Spirostomum* at pH 7·8, while others are similarly sensitive to acid waters with low pH.

Much research is needed before it is possible to say how far the presence of creatures in any particular water is solely dependent on the pH, but obviously, from the examples which have been quoted, it seems certain that the variations over the whole season must be of importance in the distribution of some animals.

There are two methods of determining the pH value of water. The electrical method requires equipment outside the reach of the student, and is rarely used outside well-equipped laboratories. The colorimetric method, however, is well within the capability of the student, and is carried out (in the field if need be) by adding a small but definite amount of a dye, called an indicator, to the liquid to be tested. The indicator changes colour in accordance with the acidity or alkalinity, and the resulting colour is then compared with those of a series of tubes containing the identical quantity of indicator to which have been added buffer solutions of known pH, so that they range widely in colour over the series.

It is unlikely that the average student will require to assess pH values more than approximately, and a very suitable indicator for such tests is the Universal Indicator supplied by British Drug Houses Ltd. This has a wide pH range, from 3 to 11, the colours varying from red at pH 3 through orange, yellow, green, blue to violet at 11. A 'capillator' is also supplied by the firm, consisting of a card on which are fixed a series of sealed capillary tubes containing the indicator and buffer solutions to give the full range of coloured tubes, each of which has the pH printed against it. Thus by comparing the water being tested, now coloured by the indicator, against the tubes of the capillator, the exact colour can be matched and the pH quickly determined. For more accurate determinations, an indicator can then be selected with a narrower range which corresponds to the approximate pH found by the Universal Indicator and preferably used with a Lovibond Comparator.

ANGLERS' NAMES FOR AQUATIC INSECTS

Many species of aquatic insects have been imitated in artificial fly-fishing and have thus received popular names. Some of the best known of these are given below, but a few of the popular names are somewhat loosely applied and may refer to more than one species, so that too much reliance should not be placed on them in identification.

Stoneflies

The Stonefly	*Dinocras cephalotes; Perla bipunctata*
The Creeper	Nymph of either of the above two species
Yellow Sally	*Isoperla grammatica* and probably other species
February Red	Female of *Taeniopteryx nebulosa*
Early Brown	Several species of the family *Nemouridae*
Willow Fly	*Leuctra geniculata*
Needle Flies	Other species of *Leuctra*

Mayflies

(The sub-imago stage is usually called a 'dun' and the fully developed imago a 'spinner'.)

The Mayfly	*Ephemera vulgata; E. danica*
Green Drake	Sub-imago of *E. danica*
Grey Drake	Imago (female) of *E. danica*
Black Drake	Imago (male) of *E. danica*
Spent Gnat	Imago of *E. danica* floating on the surface after oviposition
Claret Dun	*Leptophlebia vespertina sub-imago*
Turkey Brown	Sub-imago of *Paraleptophlebia sub-marginata*
Blue-winged Olive	Sub-imago of *Ephemerella ignita*
Sherry Spinner	Imago of *E. ignita*
White Midge	Species of *Caenis*
Angler's Curse	Species of *Caenis*
Pale Watery Duns	*Baëtis bioculatus; Centroptilum luteolum; Procloeon rufulum*
Large Dark Olive	*Baëtis rhödani*
Red Spinner	Imago of *B. rhödani* (female)
Medium Olive	Sub-imago of *B. vernus*
Olive Duns	*B. atrebatinus; B. tenax; B. rhödani; B. vernus; B. scambus* (Sub-imagines)
Iron Blue Dun	*B. pumilus; B. niger* (Sub-imagines)
Jenny Spinner	Male imago of *B. pumilus* and *B. niger*

Yellow Upright	Male imago of *Rhithrogena semicolorata*
True March Brown	Sub-imago of *R. haarupi*
Great Red Spinner	*R. haarupi; Ecdyonurus venosus; E. dispar; E. torrentis*
False March Brown	Sub-imago of *E. venosus*
August or Autumn Dun	Sub-imago of *E. dispar*
Little Yellow May Dun	*Heptagenia sulphurea*
Yellow Hawk	Sub-imago of *H. sulphurea*

Caddis-flies

Large Red Sedge	*Phryganea grandis; P. striata*
Cinnamon Sedge	*Limnophilus lunatus*
Brown Sedge	*Anabolia nervosa*
Welshman's Button	*Sericostoma personatum*
Grey Sedge	*Odontocerum albicorne*
Black Silver Horn	*Leptocerus aterrimus*
Brown Silver Horn	Other species of *Leptocerus*
Black Silver Horns	Species of *Mystacides*
Grouse Wing	*M. longicornis*
Grannom or Green-tail	*Brachycentrus subnubilus*

BIBLIOGRAPHY

The books listed below will be found helpful for further reading on the groups or subjects treated in preceding chapters. The list is by no means exhaustive, but is offered purely as a personal selection of works which the author has himself found useful.

Those marked with an asterisk are American works.

General Works

Borradaile, L. A., and Potts, F. A., 1963. *The Invertebrata*. 4th Edition. Cambridge University Press.

Brown, E. S., 1955. *Life in Fresh Water*. Oxford University Press.

*Bucksbaum, R., 1948. *Animals Without Backbones*. Chicago University Press.

Carpenter, K. E., 1928. *Life in Inland Waters*. Sidgwick & Jackson.

Clegg, J., 1967. *The Observer's Book of Pond Life*. 2nd Edition. Warne.

Furneaux, W., 1935. *Life in Ponds and Streams*. New impression. Longmans, Green.

Gardiner, J. S. (Editor), 1925–32. *The Natural History of Wicken Fen*. Bowes & Bowes (Cambridge).

*Hutchinson, G. E., 1966. A Treatise on Limnology. Vol. 1 Geography, Physics and Chemistry; Vol. 2 Introduction to Lake Biology: the Plankton. John Wiley & Sons.

Lulham, R., 1937. *An Introduction to Zoology through Nature Study*. Macmillan.

Macan, T. T., 1959. *A Guide to Freshwater Invertebrate Animals*. Longman.

Macan, T. T., 1963. *Freshwater Ecology*. Longman.

Macan, T. T., 1970. *Biological Studies of the English Lakes*. Longman.

Macan, T. T., 1973. *Ponds and Lakes*. Allen & Unwin.

Macan, T. T., and Worthington, E. B. 1973. *Life in Lakes and Rivers*. 2nd Edition. Collins.

Mellanby, H., 1963. *Animal Life in Fresh Water*. 6th Edition. Methuen.

*Morgan, A. H., 1930. *Field Book of Ponds and Streams*. Putnam, New York.

*Needham, J. G., and Lloyd, J. T., 1937. *Life of Inland Waters*. Comstock (Constable & Co. Ltd.).

*Needham, J. G., and Needham, P. R., 1962. *A Guide to the Study of Freshwater Biology*. 5th Edition. (Mainly illustrated keys, but of American forms.) Comstock (Constable & Co. Ltd.).

Popham, E. J., 1961. *Some Aspects of Life in Fresh Water*. 2nd Edition. Heinemann.

*Reid, G. K., 1961. *Ecology of Inland Waters and Estuaries*. Reinhold Publishing (Chapman & Hall).

Ruttner, F., 1953. *Fundamentals of Limnology*. University of Toronto Press.

*Ward, H. B., and Whipple, G. C., 1959. *Freshwater Biology*. 2nd Edition. Wiley New York (Chapman & Hall).

*Welch, P. S., 1962. *Limnology*. Revised Edition. McGraw Hill, New York.

Aquatic Plants

Arber, A., 1920. *Water Plants*. Cambridge University Press.

Bursche, E. M., 1971. *A Handbook of Water Plants*. Warne.

Clapham, A. R., Tutin, T. G., and Warburg, E. F., 1962. *Flora of the British Isles*. 2nd Edition. Cambridge University Press.

Fritsch, F. E., 1935. *Structure and Reproduction of the Algae*, Vol. I. C.U.P.

Fritsch, F. E., 1945. *Structure and Reproduction of the Algae*, Vol. II. C.U.P.

Round, F. E., 1965. *The Biology of the Algae*. Edward Arnold.

Sculthorpe, C. D., 1967. *The Biology of Aquatic Vascular Plants*. Edward Arnold.

West, G. S. and Fritsch, F. E., 1927. *A Treatise on the British Freshwater Algae*. Cambridge University Press.

Williams, I. A., 1946. *Flowers of Marsh and Stream*. King Penguin.

Microscopical Organisms Generally

Cash, J., and Wailes, C. H., 1904–21. *The British Freshwater Rhizopoda and Helizoa*. Vols. 1–5. Ray Society, London.

Garnett, W. J., 1953. *Freshwater Microscopy*. Constable & Co. Ltd.

Kent, W. Saville, 1880. *A Manual of the Infusoria*. Vols. 1–3. London.

Plaskett, F. J. W., 1926. *Microscopic Fresh Water Life*. Chapman & Hall.

Shipley, A. E., 1928. *Hunting under the Microscope*. Benn, London.

*Stokes, A. C., 1918. *Aquatic Microscopy*. John Wiley & Sons (Chapman & Hall).

Flatworms

Dawes, B., 1946. *The Trematoda*. Cambridge University Press.

Dawes, B., 1947. *The Trematoda of British Fishes*. Ray Society, London.

Rotifers

Donner, J., 1966. *Rotifers*. English translation by H. G. S. Wright. Warne.

Galliford, A. L., 1961–2. *How to Begin the Study of Rotifers*. *Countryside*, Vol. XIX. Dorking.

*Harring, H. K., 1922. *The Rotifera of Wisconsin*. Wisconsin Academy of Sciences, Arts and Letters.

*Harring, H. K., 1913. *Synopsis of the Rotatoria*. U.S. National Museum Bulletin No. 51.

(These two works incorporate the changes in the nomenclature of this group.)

Hollowday, E. D., 1945–50. 'Introduction to the Study of the Rotifera'. *The Microscope*. Vols. V, VI, VII. London.

Polyzoa

Allman, G. J., 1856. *Monograph of the Freshwater Polyzoa*. Ray Society, London. (Although published in 1856 it still remains the best account of the freshwater Polyzoa in English, although the classification is out of date.)

Harmer, S. F., 1922. *Cambridge Natural History*, Vol. II. Chapters XVII–XIX, 'Polyzoa'. Macmillan.

Annelids

Brinkhurst, R. O., 1971. *A Guide for the Identification of British Aquatic Oligochaeta*. Freshwater Biological Assoc., Ambleside. Scientific Publication No. 22.

Mann, K. H., 1964. *A Key to the British Freshwater Leeches.* Freshwater Biological Association Scientific Publication No. 14.

Mann, K. H., 1962. *Leeches (Hirudinea).* Pergamon.

Crustacea

Calman, W. T., 1911. *Life of Crustacea.* Methuen.

Green, J., 1961. *A Biology of Crustacea.* Witherby, London.

Gurney, R., 1931–3. *British Freshwater Copepoda,* Vols. I, II, III. Ray Society, London.

Harding, J. P. and Smith, W. A., 1960. *A Key to the British Freshwater Cyclopid and Calanoid Copepods.* Freshwater Biological Association, Ambleside. Scientific Publication No. 18.

Hynes, H. B. N., Macan, T. T., and Williams, W. D., 1960. *A Key to the British Species of Crustacea: Malacostraca.* Freshwater Biological Association, Ambleside. Scientific Publication No. 19.

Reid, D. M., 1944. *Key to the Families of British Gammaridea.* Linnean Society. Synopses of the British Fauna No. 3. London.

Scourfield, D. J., and Harding, J. P., 1958. *Key to the British Species of Freshwater Cladocera with notes on their Ecology.* 2nd Edition. Freshwater Biological Association, Ambleside. Scientific Publication No. 5.

Insects—General

Miall, L. C., 1902. *The Natural History of Aquatic Insects.* Macmillan.

Mosley, M. E., 1936. *Insect Life and the Management of a Trout Fishery.* Routledge.

Stoneflies

Hynes, H. B. N., 1958. *A Key to the Adults and Nymphs of British Stoneflies (Plecoptera).* Freshwater Biological Association. Scientific Publication No. 17.

Mayflies

Kimmins, D. E., 1972. *A Revised Key to the Adults of the British Species of Ephemeroptera.* Freshwater Biological Association. Scientific Publication No. 15.

Macan, T. T., 1970. *A Key to the Nymphs of British Ephemeroptera.* Freshwater Biological Association. Scientific Publication No. 20.

Dragonflies

Corbet, P. S., Longfield, C., and Moore, N. W., 1960. *Dragonflies.* Collins.

Longfield, C., 1949. *The Dragonflies of the British Isles.* Warne.

Lucas, W. J., 1900. *British Dragonflies.* L. Upcott Gill, London.

Lucas, W. J., 1930. *The Aquatic (Naiad) Stage of the British Dragonflies.* Ray Society, London.

Water Bugs

Macan, T. T., 1956. *A Key to the British Water Bugs.* Freshwater Biological Association. Scientific Publication No. 16.

Southwood, T. R. E., and Leston, D., 1959. *Land and Water Bugs of the British Isles.* Warne.

Alder-flies, etc.

Kimmins, D. E., 1944. *Key to the British Species of Aquatic Megaloptera and Neuroptera*. Freshwater Biological Association. Scientific Publication No. 8.

Caddis-flies

Hickin, N. E., 1952. *Caddis*. Methuen.

Hickin, N. E., 1946. 'Larvae of the British Trichoptera', *Transactions of the Royal Entomological Society*, London. Vol. 97, Part 8. (1946 and subsequent transactions.)

Hickin, N. E., 1967. *Larvae of the British Trichoptera*. Hutchinson.

Mosley, M. E., 1939. *The British Caddis Flies*. Routledge.

Aquatic Moths

Beirne, B. P., 1954. *British Pyralid and Plume Moths*. Warne.

Beetles

Balfour-Browne, F., 1940–1958. *British Water Beetles*. Vols. I, II and III. Ray Society, London.

Balfour-Browne, F., 1925. *Concerning the Habits of Insects*. Cambridge University Press.

Holland, D. G., 1972. *A Key to the larvae, pupae and adults of the British species of Elminthidae*. Freshwater Biological Association. Scientific Publication No. 26.

Linssen, E. F., 1959, *Beetles of the British Isles*. 2 vols. Warne.

Two-winged Flies

Collyer, C. N. and Hammond, C. O., 1968. *Flies of the British Isles*. 2nd Edition. Warne.

Davies, Lewis, 1968. *A Key to the British Species of Simulidae*. Freshwater Biological Association. Scientific Publication No. 24.

Edwards, F. W., 1931. *Mosquitoes and their relation to Disease*. British Museum (Natural History). Economic Series No. 4.

Edwards, F. W., and James, S. P., 1934. *British Mosquitoes and their Control*. British Museum (Natural History). Economic Series No. 4A.

Marshall, J. F., 1938. *The British Mosquitoes*. British Museum (Natural History).

Smart, J., 1968. *The British Simulidae with Keys to the Species in the Adult, Pupal and Larval Stages*. Freshwater Biological Association. Scientific Publication No. 9.

Arachnids

Bristowe, W. S., 1958. *The World of Spiders*. Collins.

Hopkins, C. L., 1961. A Key to the Water Mites of the Flatford area. *Field Studies*, Vol. 1, No. 3.

Locket, G. H., and Millidge, A. F., 1951–3. *British Spiders*. 2 vols. Ray Society, London.

Soar, C. D., and Williamson, W., 1925–29. *British Hydracarina (Water Mites)*, Vols. I–III. Ray Society, London.

T

Tardigrades

Le Gros, A., 1958. How to Begin the Study of Tardigrades. *Countryside.* Vol. XVIII, No. 8. Dorking.

Molluscs

Boycott, A. E., 1936. The Habitats of Freshwater Mollusca in Britain, *Journal of Animal Ecology*, Vol. 5, No. 1. London.

Ellis, A. E., 1925. *British Snails.* Oxford University Press.

Ellis, A. E., 1946. *Freshwater Bivalves (Corbicula, Sphaerium, Dreissena).* Linnean Society of London Synopses of the British Fauna No. 4.

Ellis, A. E., 1947. *Freshwater Bivalves (Unionacea).* Linnean Society of London Synopses of the British Fauna No. 5.

Fretter, V., and Graham, A., 1962. *British Prosobranch Molluscs.* Ray Society, London.

Macan, T. T., 1960. *A Key to the British Fresh and Brackish Water Gastropods.* 2nd Edition. Freshwater Biological Association. Scientific Publication No. 13.

McMillan, N. F., 1968. *British Shells.* Warne.

Vertebrates

Bagenal, T. B., 1970. *The Observer's Book of Freshwater Fishes.* Revised Edition. Warne.

Coward, T. A., 1969. *Birds of the British Isles.* Revised Edition, Edited by J. A. G. Barnes. Warne.

Jenkins, J. Travis, 1936. *The Fishes of the British Isles.* 2nd Edition. Warne.

Sandars, Edmund, 1937. *A Beast Book for the Pocket.* Oxford University Press.

Smith, Malcolm. *British Amphibians and Reptiles.* Collins.

Use of the Microscope

Johnson, Jean C., 1935. *Microscopic Objects. How to Mount Them.* English Universities Press.

Martin, L. C., and Johnson, B. K., 1931. *Practical Microscopy.* Blackie.

Olliver, C. W., 1947. *The Intelligent Use of the Microscope.* Chapman & Hall.

Wells, A. L., 1969. *The Microscope Made Easy.* 2nd Edition. Warne.

Wright, Lewis N. D. *A Popular Handbook to the Microscope.* Religious Tract Society.

Applications of Freshwater Biology

Hall, C. B., 1936. *The Culture of Fish in Ponds.* H.M. Stationery Office. (Ministry of Agriculture and Fisheries Bulletin No. 12.)

Hastings, A. B., 1948. *Biology of Water Supply.* British Museum (Natural History). Economic Series No. 7A.

Hynes, H. B. N., 1960. *The Biology of Polluted Waters.* Liverpool University Press.

Pearsall, Gardiner and Greenshields, 1946. *Freshwater Biology and Water Supply in Britain.* Freshwater Biological Association. Scientific Publication No. 11.

Tomlinson, T. C., 1946. *Animal Life in Percolating Filters.* Water Pollution Research Board. (Department of Scientific and Industrial Research.) Technical Paper No. 9.

*Whipple, G. C., 1914. *The Microscopy of Drinking Water.* John Wiley & Sons, New York.

INDEX